# Basic Science and Electronics

## Electrical Installation Series – Foundation Course

Ted Stocks
Charles Duncan
Malcolm Doughton
Ron Wilcox

Edited by Chris Cox

First published 1998 by
MACMILLAN PRESS LTD
Houndmills, Basingstoke, Hampshire RG21 6XS
and London
Companies and representatives throughout the world

ISBN 0–333–71987–5

A catalogue record for this book is available from the British Library.

This book is printed on paper suitable for recycling and made from fully managed and sustained forest sources.

10 9 8 7 6 5 4 3 2 1
07 06 05 04 03 02 01 00 99 98

Printed in Great Britain by L&S Printing Co. Ltd

# About this book

"Basic Science and Electronics" is one of a series of books published by Macmillan Press Ltd related to Electrical Installation Work. The series may be used to form part of a recognised course, for example City and Guilds Course 2360, or individual books can be used to update knowledge within particular subject areas. A complete list of titles in the series is given below.

Foundation Course books give the student the underpinning knowledge criteria required for City and Guilds Course 2360 Part I Theory. The supplementary book, Practical Requirements and Exercises, covers the additional underpinning knowledge required for the Part I Practice.

## *Level 2 NVQ*

Candidates who successfully complete assignments towards the City and Guilds 2360 Theory and/or Practice Part I can apply this success towards Level 2 NVQ through a process of Accreditation of Prior Learning.

# Electrical Installation Series

### *Foundation Course*

Starting Work
Procedures
Basic Science and Electronics

**Supplementary title:**
Practical Requirements and Exercises

### *Intermediate Course*

The Importance of Quality
Stage 1 Design
Intermediate Science and Theory

**Supplementary title:**
Practical Tasks

### *Advanced Course*

Advanced Science
Stage 2 Design
Electrical Machines
Lighting Systems
Supplying Installations

# Acknowledgements

The authors and publishers would like to thank the following for their help in producing this book:

Farnell for illustrations 5.1, 5.34, 7.27 and the photograph on p. 96

Maplin Electronics for illustrations 7.1, 7.28 and the photographs on pp. 94 and 97

RS Components for all other photographic illustrations of components

# Study guide

This studybook has been written to enable you to study either in a classroom or in an open or distance learning situation. To ensure that you gain the maximum benefit from the material you will find prompts all the way through that are designed to keep you involved with the subject. The book has been divided into 24 parts each of which may be suitable as one lesson in the classroom situation. However if you are studying by yourself the following points may help you.

- ☞ Work out when, and for how long, you can study each week. Complete the table below and from this produce a programme so that you will know approximately when you should complete each chapter and take the progress and end tests. Your tutor may be able to help you with this. It may be necessary to reassess this timetable from time to time according to your situation.
- ☞ Try not to take on too much studying at a time. Limit yourself to between 1 hour and 2 hours and finish with a task or the self assessment questions (SAQ). When you resume your study go over this same piece of work before you start a new topic.
- ☞ You will find the answers to the questions at the back of this book but before you look at the answers check that you have read and understood the question and written the answer you intended.
- ☞ A "progress check" at the end of the Science Section and an "end test" covering all the material in this book are included so that you can assess your progress.
- ☞ Tasks are included where you are given the opportunity to ask colleagues at work or your tutor at college questions about practical aspects of the subject. There are also tasks where you may be required to use manufacturers' catalogues to look answers up. These are all important and will aid your understanding of the subject.
- ☞ Your safety is of paramount importance. You are expected to adhere at all times to current regulations, recommendations and guidelines for health and safety.

**Study times**

| | a.m. from | to | p.m. from | to | Total |
|---|---|---|---|---|---|
| Monday | | | | | |
| Tuesday | | | | | |
| Wednesday | | | | | |
| Thursday | | | | | |
| Friday | | | | | |
| Saturday | | | | | |
| Sunday | | | | | |

| **Programme** | Date to be achieved by |
|---|---|
| Chapter 1 | |
| Chapter 2 | |
| Chapter 3 | |
| Chapter 4 | |
| Progress check | |
| Chapter 5 | |
| Chapter 6 | |
| Chapter 7 | |
| Chapter 8 | |
| End test | |

# Contents

# 1

# The Basics

At the beginning of all the other chapters in this book you will be asked to complete a revision exercise based on the previous chapter – Sid holding a clipboard will remind you of this. For the first exercise, see what you can find out about the following scientists and which SI units (International System of Units) were named after them. (A good dictionary could help!)

André-Marie Ampère
SI Unit: ______

Sir Isaac Newton
______

James Prescott Joule
______

James Watt
______

Blaise Pascal
______

Count Alessandro Giuseppe Antonio Anastasio Volta
______

Georg Simon Ohm
______

Ernst Werner von Siemens
______

**On completion of this chapter you should be able to:**

- state the basic principles, units and derived units of the SI system
- name the symbols and factors appropriate to multiples and submultiples of SI units
- perform calculations including the transposition and use of formulae
- complete calculations using a calculator
- complete the revision exercise at the beginning of the next chapter

# Part 1

## SI units

The International System of Units (SI)

When Noah built his ark his measurements were in cubits – that was the distance between his elbow and the end of his fingers. Some of the other measurements that were used then were palm, span and stride. Because these measurements were approximate and depended on the size of the person using them it became obvious that "standard" measurements were needed (Figure 1.1). These standards were never to vary so that if, for example, one was selling beer then 1 pint was the same for one customer as another.

In different countries different standards were set.

In Britain we used miles, furlongs, yards, feet and inches for measuring length and distance. In Europe kilometres, metres, centimetres and millimetres were used. Although we also now use the metric system there are instances where we still use Imperial Measures.

*Figure 1.1 Before measurements were standardized!*

# The use of SI units

The SI unit for length is the metre. The unit's symbol is m and the quantity symbol (as used in equations) is $l$. There are seven base and two supplementary units in The International System of Units. The base units you need to know about at this stage are shown in Table 1.1.

*Table 1.1 Base units*

| | Unit | Unit symbol | Quantity symbol |
|---|---|---|---|
| Length | metre | m | $l$ |
| Mass | kilogram | kg | $m$ |
| Time | second | s | $t$ |
| Electric current | ampere | A | $I$ |
| Temperature* | kelvin | K | $t$ |

*Although the SI unit is the kelvin it has been internationally agreed that the **degree Celsius (°C)** will be the unit for everyday temperature measurement. The kelvin starts at absolute zero; it is the fraction 1/273.15 of the thermodynamic temperature of the triple point of water. The degree Celsius is the kelvin minus 273.15, so 0 °C corresponds to 273.15 K. The intervals are the same for both.

## Derived units

Other units are derived from the base units.

**The unit of force, the newton** (unit symbol N, quantity symbol $F$), is mass multiplied by acceleration. (Acceleration is speed divided by time and speed is distance divided by time.)
1 N = 1 kg m/s$^2$

A **weber** (Unit symbol Wb, quantity symbol $F$) is the unit of magnetic flux, causing the electromotive force of one volt in a circuit of one turn when generated in one second.
A single conductor passing through a magnetic flux at the rate of 1 Wb/s generates 1 volt.

The **tesla** (Unit symbol T, quantity symbol $B$) is the unit of magnetic flux density.
1 T = 1 Wb/m$^2$

## Other commonly used units

**Area** is the product of two lengths multiplied together.
As length is measured in metres then area would be in square metres: m$^2$.

**Speed** (velocity) is a distance divided by time
Distance is measured in metres and time in seconds so the speed is in metres per second: m/s.

Symbols for some common quantities are shown in Table 1.2.

*Table 1.2 Some common quantity symbols*

| Quantity | Unit | Unit symbol | Quantity symbol |
|---|---|---|---|
| acceleration | metre/second$^2$ | m/s$^2$ | $a$ |
| area (cross sectional area) | square metre | m$^2$ | $A$ |
| capacitance | farad | F | $C$ |
| charge (quantity of electricity) | coulomb | C | $Q$ |
| density | kilogram/cubic metre | kg/m$^3$ | ρ |
| electromotive force (e.m.f.) | volt | V | $V$ |
| electrical energy | joule (kilowatt hour | J kWh) | $W$ |
| force | newton | N | $F$ |
| frequency | hertz | Hz | $f$ |
| inductance | henry | H | $L$ |
| potential difference | volt | V | $V$ |
| power | watt | W | $P$ |
| resistance | ohm | Ω | $R$ |
| resistivity | ohm metre | Ωm | ρ |
| speed (velocity) | metre/sec | m/s | $v$ |
| volume | cubic metre | m$^3$ | $V$ |
| work | newton metre | Nm | $W$ |
| magnetic flux | weber | Wb | Φ |
| magnetic flux density | tesla | T | $B$ |

## Multiples and submultiples of units

Sometimes it is more convenient to use multiples and submultiples of units (Table 1.3).

*Table 1.3*

| **Prefix** | **Symbol** | **Value** | |
|---|---|---|---|
| tera | T | $10^{12}$ or | 1 000 000 000 000 |
| giga | G | $10^9$ | 1 000 000 000 |
| mega | M | $10^6$ | 1 000 000 |
| kilo | k | $10^3$ | 1 000 |
| deci | d | $10^{-1}$ | 0.1 |
| centi | c | $10^{-2}$ | 0.01 |
| milli | m | $10^{-3}$ | 0.001 |
| micro | μ | $10^{-6}$ | 0.000001 |
| nano | n | $10^{-9}$ | 0.000000001 |
| pico | p | $10^{-12}$ | 0.000000000001 |

**Powers of ten**

In the above table $10^3$ should be read as "ten to the power of three" and $10^{-3}$ as "ten to the power of minus three", which is another way of putting $1/10^3$. "The power of ten" means how many times 1 is multiplied by 10, so $10^3$ is the shorthand way of writing $1 \times 10 \times 10 \times 10$ or 1000.
Note that $10^1$ is $1 \times 10 = 10$ and $10^0 = 1$.

The most commonly used powers of ten in electrical units are the:

**megawatt**

1 megawatt = 1 000 000 watts
1 MW = $10^6$ W
or 6.4 MW = 6 400 000 watts or $6.4 \times 10^6$ watts

**kilowatt**

1 kilowatt = 1000 watts
1 kW = $10^3$ W
or 4.7 kW = 4700 watts or $4.7 \times 10^3$ watts

**milliampere**
1 ampere = 1000 milliamperes
1 A = 1000 mA
1 mA = $10^{-3}$ A
16 mA is 0.016 amperes or $16 \times 10^{-3}$ amperes

**microfarad**
1 farad = 1 000 000 microfarads
1 F = 1 000 000 μF
1 μF = $10^{-6}$ F
58 μF is 0.000 058 farads or $58 \times 10^{-6}$ farads

**NOTE:**
Unit symbols are the same for singular and plural. The letter s must never be added to form a plural (ms is milliseconds *not* metres).

Units that are named after famous scientists use a capital letter for the unit symbol, for example ampere (A), farad (F), newton (N), joule (J).

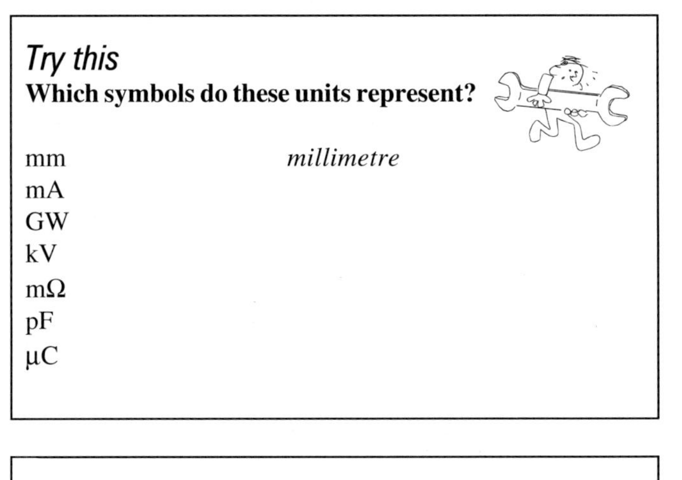

*Try this*
**Which symbols do these units represent?**

| | |
|---|---|
| mm | *millimetre* |
| mA | |
| GW | |
| kV | |
| mΩ | |
| pF | |
| μC | |

*Remember*
**A quantity has both a number and a unit.**

# Area and perimeters

## Area

The measurement of area is expressed in $m^2$ (square metres) and the common submultiples are $mm^2$, $dm^2$ and $cm^2$. In a **rectangular shape** the area is found by multiplying the length by the breadth (Figure 1.2).

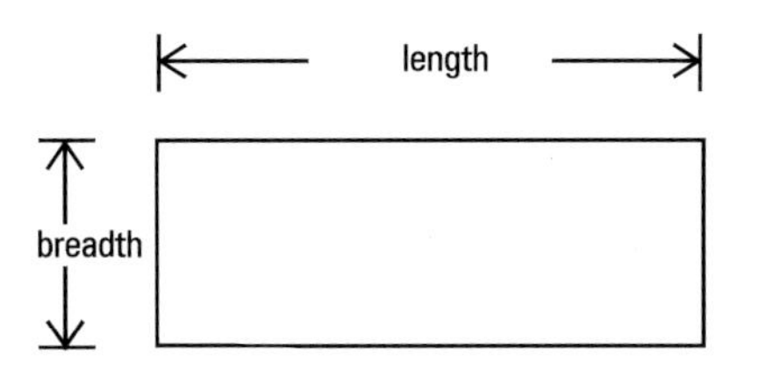


*Figure 1.2*

*Example:*
Calculate the total cross-sectional area (c.s.a.) of trunking whose dimensions are 70 mm × 50 mm.

$$\begin{aligned} \text{area} &= \text{length} \times \text{breadth} \\ &= 70 \times 50 \text{ mm}^2 \\ &= 3500 \text{ mm}^2 \end{aligned}$$

The area of other shapes is calculated by using appropriate formulae.

The area of a **circle** is calculated by the formula:

$$\text{area} = \pi r^2 \quad \text{or} \quad \frac{\pi d^2}{4}$$

where $r$ is the radius of the circle, $d$ is the diameter of the circle (and when a number is squared it is multiplied by itself) and π (pi) has a value of approximately 22⁄7 or 3.142 (Figure 1.3).

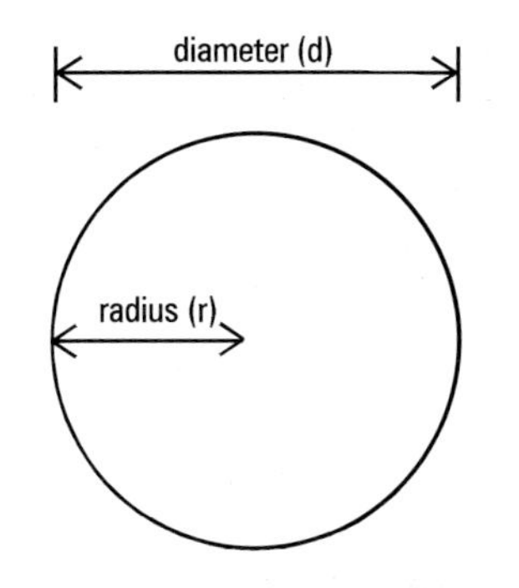


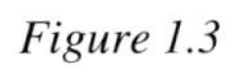

*Figure 1.3*

*Example:*
What is the area of a circle of diameter 6 m?

$$\begin{aligned} \text{area} &= \pi r^2 \\ &= \frac{22 \times 3 \times 3}{7} \\ &= 28.3 \text{ m}^2 \end{aligned}$$

The area of a **triangle** is calculated by the formula

$$\text{area} = \frac{1}{2} \text{base} \times \text{height}$$

Note that the height is the perpendicular height (Figure 1.4).

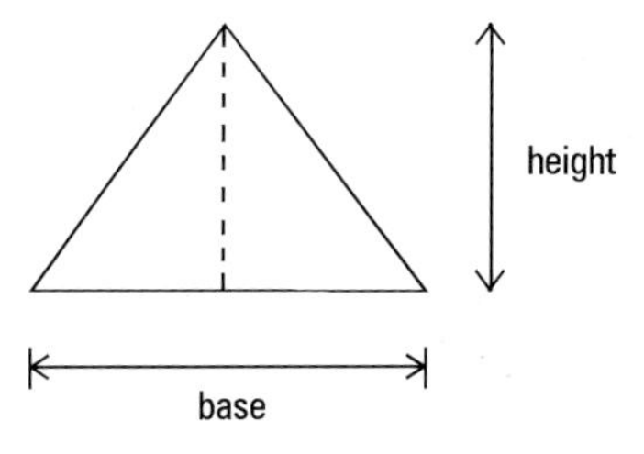


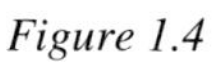

*Figure 1.4*

## Example

What is the area in $cm^2$ of a triangle with a base of 2 cm and a height of 4 cm?

$$\text{area} = \frac{1}{2}(2 \times 4)$$
$$= 4 \text{ cm}^2$$

## Try this

1. Calculate the cross-sectional area of 75 mm × 25 mm trunking. Answer to be in $mm^2$.

2. Calculate the overall cross-sectional area of a cable of nominal overall diameter of 3.5 mm. Answer to be in $mm^2$. ($\pi = 3.142$)

## Perimeters

The perimeter of a figure is the length of the enclosed outline. In a rectangle, triangle or other straight-sided figure it is the addition of the lengths of the sides (Figure 1.5).

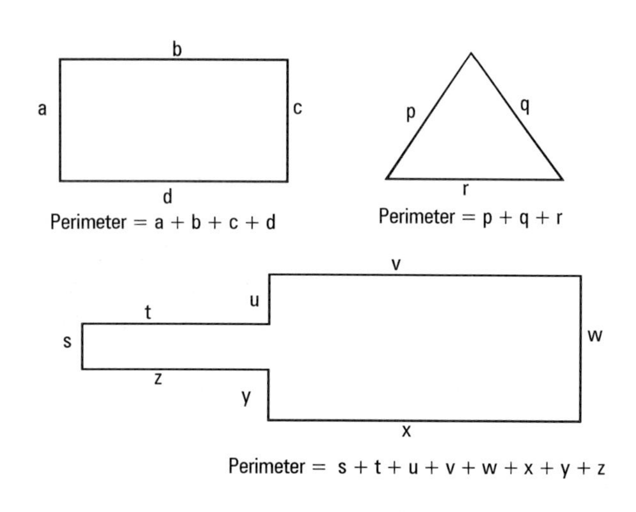

*Figure 1.5*

## Example

What is the perimeter of a square with sides of 4 cm?

$$\text{Perimeter} = \text{Side 1} + \text{Side 2} + \text{Side 3} + \text{Side 4}$$
$$= 4 + 4 + 4 + 4$$
$$= 16 \text{ cm}$$

The perimeter of a circle (called the circumference) is found by the formula

$$\text{circumference} = \pi d$$

where $d$ is the diameter of the circle (Figure 1.6).

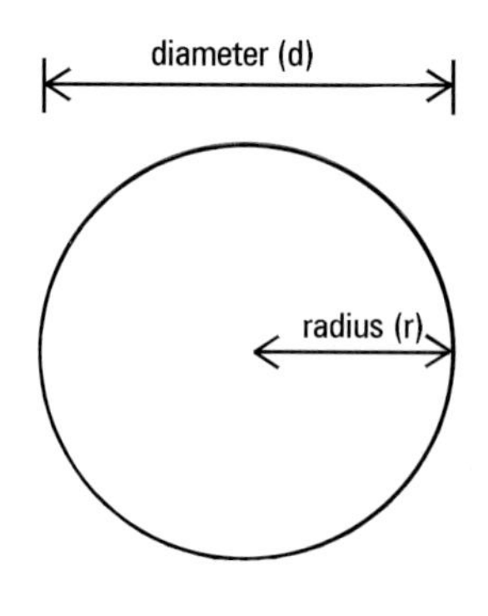


*Figure 1.6*

## Example

What is the length of the perimeter of a circle that has a radius of 2 cm? Remember that the diameter is twice the radius.

$$\text{circumference} = \pi d$$
$$= \frac{22 \times 4}{7}$$
$$= 12.57 \text{ cm}^2$$

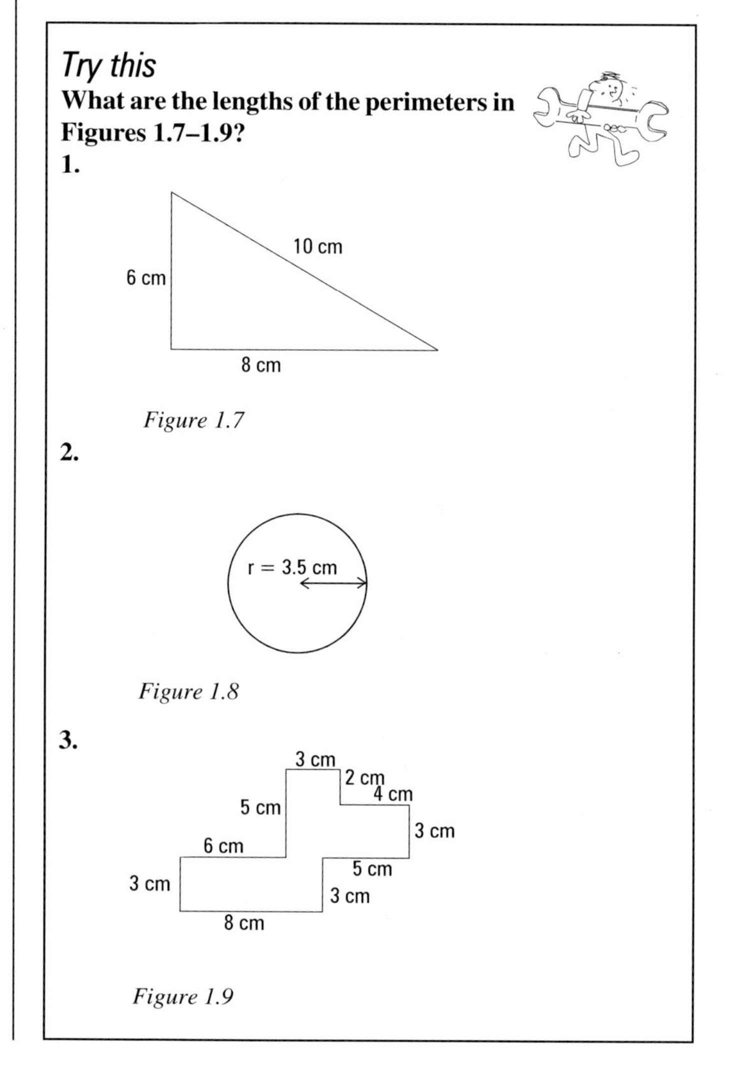

## Try this

**What are the lengths of the perimeters in Figures 1.7–1.9?**

**1.**



*Figure 1.7*

**2.**



*Figure 1.8*

**3.**



*Figure 1.9*

## Pythagoras' theorem

To find a length of one side of a triangle where one angle is a right angle we can use Pythagoras' theorem. This says that in a right-angled triangle the square on the hypotenuse is equal to the sum of the squares on the other two sides (Figure 1.10).

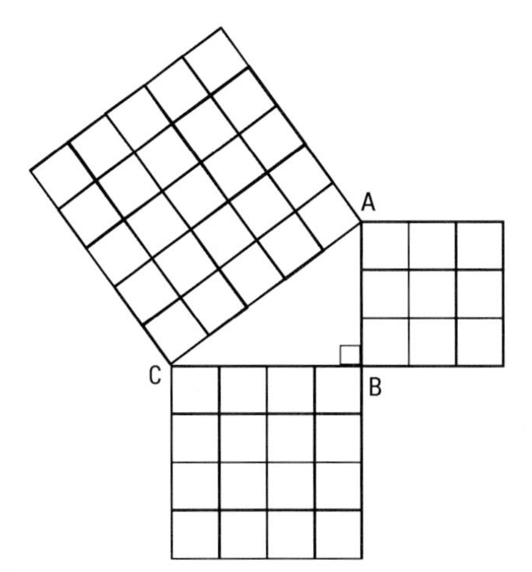


*Figure 1.10*

$$AC^2 = AB^2 + BC^2$$

$$\therefore \quad AC = \sqrt{AB^2 + BC^2}$$

Where √ is the square root sign. The square root of a specified number is the number that when multiplied by itself equals the specified number.

We can use this formula to find out the length of a ladder which is required in order to follow the 4:1 rule. Remember the 4:1 rule means that the height of a ladder above the ground must be 4 times the distance the ladder is out from the foot of the wall.

### *Example*

A ladder is placed 0.75 m out from the wall (Figure 1.11).

What length of ladder is required in order to follow the 4:1 rule?

$BC = 0.75$ so $AB \quad = 4 \times 0.75$
$\qquad\qquad\qquad\quad = 3$ m

So $\quad AC^2 = AB^2 + BC^2$
$\qquad\qquad\quad = 3^2 + 0.75^2$
$\qquad\qquad\quad = 9.56$

$AC = \sqrt{9 + 0.56}$
$\quad\;\; = 3.09$ m

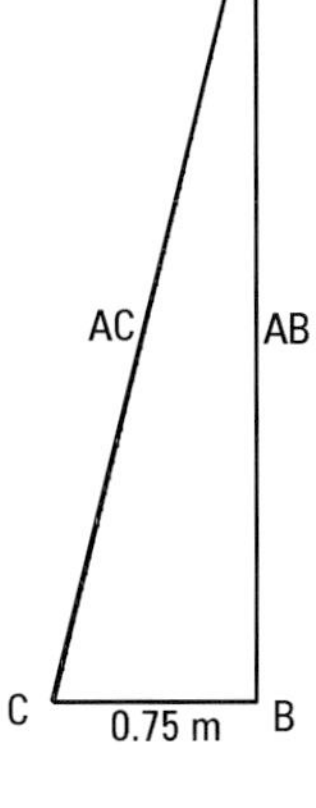


*Figure 1.11*

### *Try this*

In the right-angled triangle in Figure 1.12, find the value of the side AC.

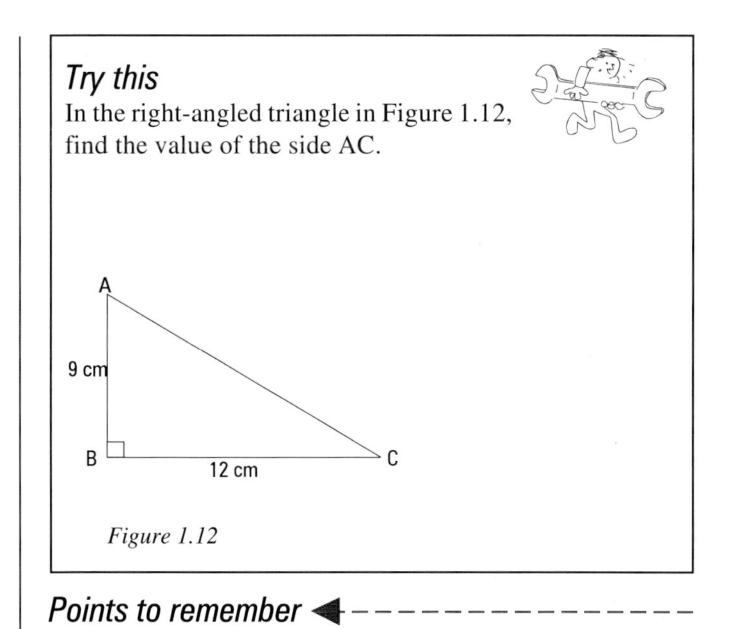


*Figure 1.12*

### *Points to remember*

## SI units

State the SI units, the unit symbol and the quantity symbol for the following. The first one is done for you.

| | **Unit** | **Unit symbol** | **Quantity symbol** |
|---|---|---|---|
| length | metre | m | *l* |
| mass | | | |
| time | | | |
| electric current | | | |
| area | | | |
| speed | | | |
| acceleration | | | |
| force | | | |
| magnetic flux | | | |
| magnetic flux density | | | |

Complete the following:

1 megawatt = 1 000 000 watts or $10^6$ W
1 kilowatt = _________ watts or __ W
1 ampere = _________ milliamperes or ____ mA
1 farad = _________ microfarads or ____ μF
1 farad = _________ nanofarads or ____ nF

A quantity has both a _________ and a _________

## Area and perimeters

The measurement of area is expressed in ____
Write down the formulae for the area of a rectangle, a circle and a triangle.

Area of rectangle =

Area of circle =

Area of triangle =

What is the value of pi?

# Part 2

This book is not intended to be a detailed mathematics book. Some students may feel that they require more help at this stage, while others may be more confident and already know what is being introduced. We would suggest to those who would like more detail and extra exercises that they consult the book "Maths for Intermediate GNVQ", also published by Macmillan Press Ltd.

## Volume

Volume is measured in $m^3$ (cubic metres). You may also find measurements in the submultiples $dm^3$, $cm^3$ and $mm^3$.

The volume of a cuboid (Figure 1.13), and in fact any regular (parallel-sided) solid is the area of one end multiplied by the length. It is found by using the formula

volume = length × breadth × depth (height)

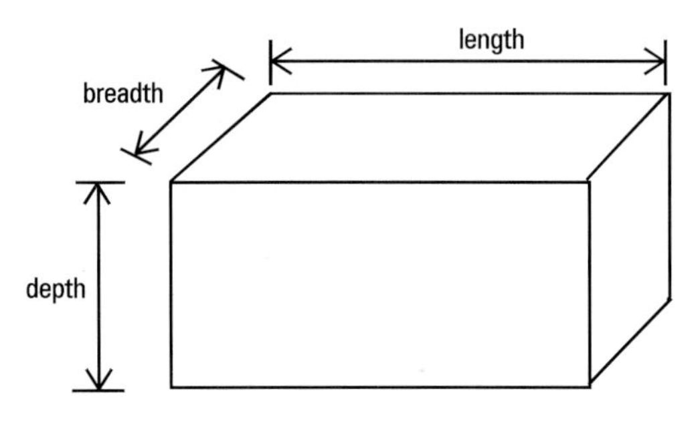


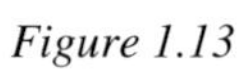
*Figure 1.13*

### *Example*

Calculate the volume of a room 6 metres × 5 metres × 2 metres.

volume = length × breadth × depth
$= 6 \times 5 \times 2$
$= 60\ m^3$

The volume of a cylindrical object is the area of the circle multiplied by the height (or length) (Figure 1.14).

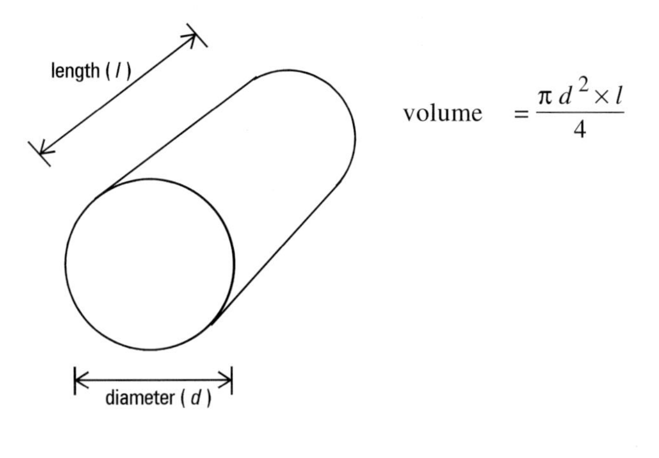


$$\text{volume} = \frac{\pi d^2 \times l}{4}$$

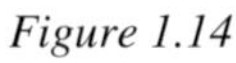
*Figure 1.14*

### *Example*

Find the volume of a cylindrical tank 2 metres high and of diameter 0.5 metres.

$$\text{volume} = \frac{3.142 \times 0.5 \times 0.5 \times 2}{4}$$
$$= 0.39\text{m}^3$$

The volume of a liquid or gas is found by the same method. Suppose you need to find out how much water a tank holds. If the tank is 1 m long, 0.75 m wide and 0.5 m high, how much water will the tank hold?
You will need to know that 1 $dm^3$ = 1 litre.
As we are working in metres the equivalent is 0.001 $m^3$ = 1 litre or 1 $m^3$ = 1000 litres.

volume of tank $= 1 \times 0.75 \times 0.5$
$= 0.375\text{m}^3$

amount of water $= 0.375 \times 1000$
= 375 litres

Can you picture the size of this tank? Is it the size of a small fish tank, the cold water tank in your loft or a swimming pool?

### *Try this*

1. Calculate the volume of a tank 3 metres by 2 metres by 1.5 metres. Answer to be in $m^3$.

2. Calculate the capacity of a cylindrical tank 0.5 m in diameter and 1.5 m high. Answer to be in $m^3$.

One way of finding the volume of an irregular figure is to find the volumes of the various parts and add them together or, as in the following example, subtract them.

### Example

What is the volume of the hollow cylinder shown in Figure 1.15? The diameter of the outside cylinder is 20 mm, the inside 15 mm and the length of the cylinder is 80 mm.

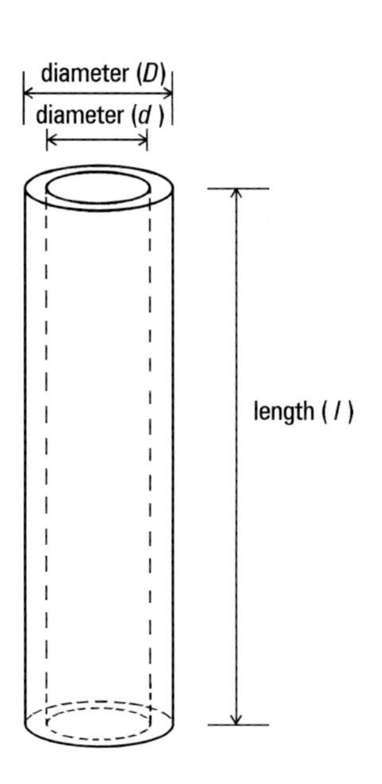


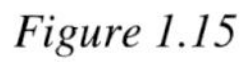

*Figure 1.15*

The volume of the cylinder if it were not hollow would be

$$\text{Volume} = \frac{\pi D^2 \times l}{4}$$

The volume of the inside cylinder is

$$\text{Volume} = \frac{\pi d^2 \times l}{4}$$

Subtracting the inside cylinder from the whole will give us the volume of the hollow cylinder:

$$\text{Volume} = \frac{\pi D^2 \times l}{4} - \frac{\pi d^2 \times l}{4}$$

$$\text{Volume} = 25\,143 - 14\,143 = 11\,000 \text{ mm}^3$$

### Example

Find the volume of the heating oil tank in Figure 1.16. The cylindrical section of the tank is 3 metres long and the tank has a diameter of 1 metre.

The shape is made up of a cylinder and two half spheres.

The formula for the volume of a sphere = $\frac{4}{3} \pi r^3$

In this formula we must use the radius of the sphere (half the diameter) not the diameter.

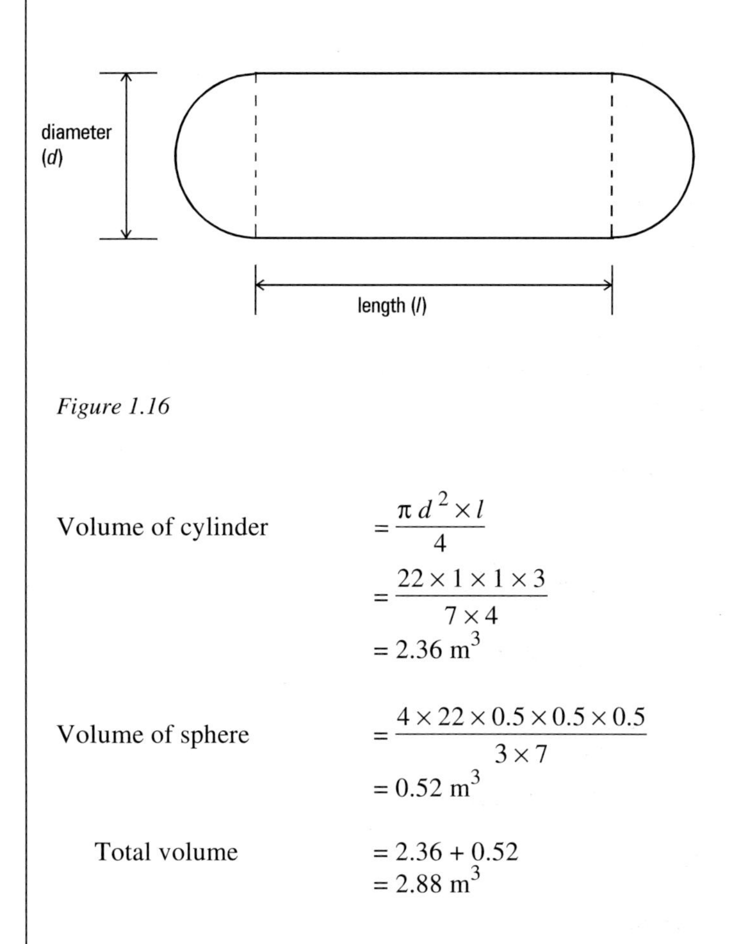


*Figure 1.16*

$$\text{Volume of cylinder} = \frac{\pi d^2 \times l}{4} = \frac{22 \times 1 \times 1 \times 3}{7 \times 4} = 2.36 \text{ m}^3$$

$$\text{Volume of sphere} = \frac{4 \times 22 \times 0.5 \times 0.5 \times 0.5}{3 \times 7} = 0.52 \text{ m}^3$$

$$\text{Total volume} = 2.36 + 0.52 = 2.88 \text{ m}^3$$

### Try this

Find the volume of the irregular shape shown in Figure 1.17.

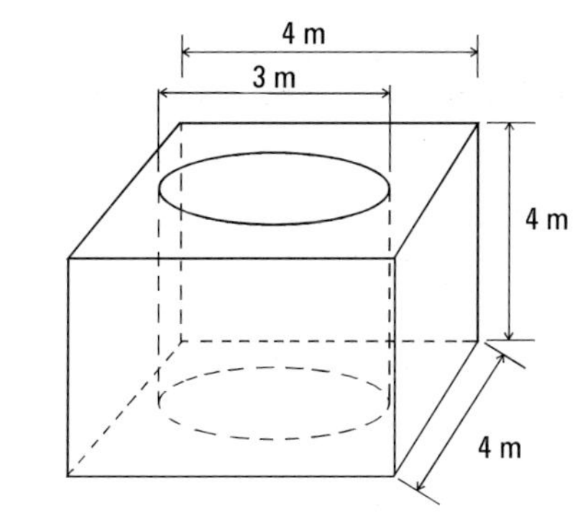


*Figure 1.17*

## Ratio, percentage and efficiency

You will probably have heard of the turns **ratio** in the context of transformers. The ratio is the proportion of turns on one side of the transformer to the turns on the other. 2:1 is a ratio – the first side is twice the second – and 1:2 is a ratio – the first side is half the second. So it is important to write down the ratio correctly or you may assume that a step-down transformer is a step-up one!

### *Example*

The windings on a transformer are in the ratio of 5:3. If the number of turns on the primary winding are 250, how many turns are on the secondary winding?

The formula for this is

$$\text{(ratio)}\ \frac{5}{3} = \frac{N_p}{N_s}$$

where $N_p$ is the number of turns on the primary winding and $N_s$ the number of turns on the secondary winding.

$$\frac{5}{3} = \frac{250}{N_s}$$

Number of turns on the secondary winding = 150.

So ratios are where one or more numerical quantities are in proportion to another (or more than one other) of the same kind of units. Ratios should be written in the simplest form possible, so 10:2 should be written as 5:1.

### *Example*

The sides of a right-angled triangle are in the ratio of 3:4:5. If the shortest side is 6 cm long, what are the lengths of the other two sides?

If the shortest side had been 3 cm long then the other sides would have been 4 cm and 5 cm long. The shortest side is twice this length, so we must also double the other two lengths to keep the ratio correct.

The lengths of the three sides of the triangle are therefore 6 cm, 8 cm and 10 cm.

A **percentage** is the proportion per hundred, i.e. 1 per cent is one part in a hundred. So 1 per cent of 200 is 2.

### *Example*

What percentage of 200 is 50?

$$\begin{aligned}\text{percentage} &= \frac{50}{200} \times 100 \\ &= 25\%\end{aligned}$$

The useful energy output of a machine is generally less than the energy input. This is because some of the energy input is wasted. In an electric machine electrical energy is converted into mechanical energy, but some of the electrical energy is converted into heat energy (friction) and noise energy and is therefore wasted energy. The **efficiency** of the machine is the ratio of the useful output energy (or power) and the input energy (or power). It is expressed by the formula:

$$\text{Efficiency} = \frac{\text{output}}{\text{input}}$$

(The symbol for efficiency is $\eta$ – eta)

Expressing this as a percentage

$$\text{Efficiency \%} = \frac{\text{output}}{\text{input}} \times 100$$

### *Example*

The input power to drive a motor is 3500 W. If the power output is 2800 W, what is the percentage efficiency of the motor?

$$\begin{aligned}\text{Efficiency \%} &= \frac{\text{output}}{\text{input}} \times 100 \\ &= \frac{2800}{3500} \times 100 \\ &= 80\%\end{aligned}$$

### *Try this*

1. The sides of a right-angled triangle are in the ratio of 5:12:13. If the shortest side is 2.5 cm long, what are the lengths of the other two sides?

2. The windings on a step-up transformer are in the ratio of 3:5. If the number of turns on the primary winding are 240, how many turns are on the secondary winding?

3. The input power to drive a motor is 3 kW. If the power output is 2.28 kW, what is the percentage efficiency of the motor?

4. If the output power of a motor is 8 kW and the input power is 10 kW, what is the percentage efficiency of the motor?

## Transposition of formulae

Very often when we are faced with solving a problem we need to use a formula to work it out. At first sight some formulae look daunting, especially when the value you need to find is in the middle of a lot of other letters! If you normally have trouble with formulae, then work carefully through the next part.

We will start by using numbers instead of letters, as it is easier to confirm that each stage is correct.

Take the addition of two loads. We can write it down as: 3 kilograms and 4 kilograms when added together make 7 kilograms.

If we carry on like this we waste a lot of time, paper and patience, so we will use the shorthand way of writing the sentence. The use of the symbols + , – , × , ÷ and = are shorthand for words.

$$3 + 4 = 7$$

This can also be written as

$$4 + 3 = 7$$

or $7 = 3 + 4$

or $7 = 4 + 3$

Using the same two loads we can also say that if we take the 3 kilogram load from the total of 7 kilograms we are left with 4 kilograms.

$$7 - 3 = 4$$

So you can see that a number which had previously been added to one side of the equation can be taken away from the other, *or*, if it was originally subtracted on one side it must be added on the other. Think of the equals sign as balancing what is on one side of the equation with what is on the other.

$3 + 4 = 7$ $\quad$ $3 = 7 - 4$

or

$3 + 4 - 7 = 0$

Now we will use letters to represent the quantities.

$$M_T = M_1 + M_2$$

and it can also be written as

$$M_1 + M_2 = M_T$$

or $M_1 = M_T - M_2$

or $M_2 = M_T - M_1$

### Example:

Resistors of 0.896 Ω ($R_1$) and 0.413 Ω ($R_2$) are connected in series.

Find the total resistance ($R_T$).

$$R_T = R_1 + R_2$$
$$R_T = 0.896 + 0.413$$
$$= 1.309\ \Omega$$

### Example:

If there are two resistors in series and one has a value of 1.275 Ω, find the other if the total is 4.6 Ω.

$$R_T = R_1 + R_2$$
$$R_2 = R_T - R_1$$
$$R_2 = 4.6 - 1.275$$
$$= 3.325\ \Omega$$

### Try this

1. Resistors of 2.45 Ω and 1.325 Ω are connected in series. Find the total resistance.

2. The total resistance in a circuit is 3.24 Ω. If one of the two resistors in series has a value of 0.868 Ω, find the value of the other resistor.

Now take the multiplication of two numbers.

If one winding of a transformer has 10 turns and the other has 5 turns, then the first has twice as many as the second

$10 = 5 \times 2$ $\quad$ or $\quad$ $5 \times 2 = 10$

or the second has only half as many as the first!

$$5 = \frac{10}{2}$$

We can also say

$$2 = \frac{10}{5}$$

Now the number that had been multiplied on one side of the equation has been divided by on the other side.

When letters are used to represent numbers the theory behind the rearranging is the same.

For example:

$$V = I R$$
$$I = \frac{V}{R}$$
$$R = \frac{V}{I}$$

This rearranging of the formula is called **transposition**.

When using letters the "multiplied by" sign is left out, but as soon as you replace the letters with numbers it must be put back in:

$$V = I R$$

$$V = 2 \times 5 \text{ volts}$$

When quantities are given to insert in a formula instead of the letters it must be remembered that the symbol in the formula represents a certain size of quantity. For example, using $V = I R$, the $V$ represents the voltage measured in volts – not kilovolts or millivolts. If the voltage is given in anything other than volts it must first be converted into the correct units. For example, 2 kV would be changed to 2000 volts before inserting into the formula.

The formulae used in this section for the purpose of covering the topic of "transposition of formulae" will be covered more fully later in this book.

## Example:

Calculate the length of copper conductor with a c.s.a. of 1.0 mm$^2$ ($1 \times 10^{-6}$ m$^2$) that gives a resistance of 0.425 Ω. Resistivity of copper $1.73 \times 10^{-8}$ Ω m

The formula to use is:

$$R = \frac{\rho\, l}{A}$$

where $R$ is the resistance in Ω
$\rho$ (rho) is the resistivity in Ω m
$l$ is the length in m
$A$ is the cross-sectional area in m$^2$

Transpose the formula by multiplying up the $A$:

$$R A = \rho\, l$$

Now divide down the $\rho$ to leave the $l$ on its own

$$\frac{R A}{\rho} = l \qquad \text{or} \qquad l = \frac{R A}{\rho}$$

$$l = \frac{0.425 \times 1 \times 10^{-6}}{1.73 \times 10^{-8}}$$

$$l = 24.57 \text{ m}$$

If we had been required to find the cross-sectional area then the formula would have been rearranged as:

$$A = \frac{\rho\, l}{R}$$

What would the formula be if we required the resistivity?

More complicated formulae are treated in the same way, but take care when you mix the operations plus and minus with multiplication and division.

For example:

$$14 - 2 = \frac{4 \times 6}{2} \qquad (12 = 12)$$

Now let's move some of the numbers around.

Remember: the equation that is already there must be kept whole if you are mixing the operations plus and minus with multiply and divide: for example, you need $(14 - 2) \times 2$, *not* $14 - 2 \times 2$.

This also applies to an item in the equation that you wish to move – when mixing operations you must arrange the equation so that the item is on its own before you move it.

$$-2 = \frac{(4 \times 6)^*}{2} - 14 \qquad \textbf{not} \qquad -2 = \frac{4 \times 6 - 14}{2}$$

$$14 = \frac{(4 \times 6)}{2} + 2 \qquad \textbf{not} \qquad 14 = \frac{4 \times 6 + 2}{2}$$

$$\frac{14 - 2}{6} = \frac{4}{2} \qquad \textbf{not} \qquad \frac{14}{6} - 2 = \frac{4}{2}$$

$$\frac{14 - 2}{4} = \frac{6}{2}$$

$$(14 - 2) \times 2 = 6 \times 4$$

## Example:

A length of copper wire has a resistance of 50 Ω at 10 °C. Calculate its resistance at 40 °C.
Temperature coefficient of resistance (symbol α – alpha) for copper is 0.004 Ω / Ω °C.

$$\frac{R_1}{R_2} = \frac{(1 + \alpha\, t_1)}{(1 + \alpha\, t_2)}$$

where $R$ is the resistance
$t$ is the temperature

$$R_1 (1 + \alpha\, t_2) = R_2 (1 + \alpha\, t_1)$$

$$R_2 = \frac{R_1 (1 + \alpha t_2)}{(1 + \alpha t_1)}$$

$$R_2 = \frac{50[1 + (0.004 \times 40)]}{1 + (0.004 \times 10)}$$

$$R_2 = 55.77\ \Omega$$

### *Try this*

1. The current in a circuit is 100 A and the resistance is 0.2 Ω. Calculate the voltage.

2. A voltage of 240 V is applied to a resistor of 80 Ω. Calculate the current which flows.

### *Points to remember*

Complete the following:

Volume is measured in ______. Submultiples that may also be used are

The volume of a cuboid (or a box) is found by using the formula:

The volume of a cylindrical object is found by using the formula:

In addition to the way of finding the volume of an irregular object that we looked at in this chapter an alternative option is to immerse it in a tank of water and calculate the displacement of the water.

Transpose the following formulae to find $R$:

$P = I^2 R$ $\qquad R =$

$P = \frac{V^2}{R}$ $\qquad R =$

$I = \frac{V}{R}$ $\qquad R =$

# Part 3

## Method of procedure for solving mathematical problems

You will have realised by now that to solve a mathematical problem may require several actions to be taken, and that those actions must follow a particular order.

1. Decide which formula is needed and write it down.
2. Transpose it if necessary.
3. Convert the given data to SI units where necessary.
4. Substitute the data information in the formula and perform the calculation.
5. Adjust the final value if it is required in a different form.

### *Example:*

A fire element has a working resistance of 12 Ω when supplied with 240 volts. Find the current flowing in the circuit.

1. Write down the formula. This formula is, as you probably know, Ohm's Law, and we will be covering this law more fully in Chapter 3.

   $V = I R$

2. Transpose it if necessary.

   $V = I R$

   $I = \frac{V}{R}$

3. Convert data to SI units. Not applicable this time.
4. Substitute the data into the formula and perform the calculation. $V = 240$ V, $R = 12\ \Omega$

   $I = \frac{240}{12}$

   $I = 20$ A

5. Adjust the final value if required in different value of units. However, that is not necessary.

### *Example:*

What is the resistance of a material with a resistivity of 17.3 μΩ mm, a cross-sectional area of 1.5 mm$^2$ and length of 5 m?

1. Write down the formula:

   $R = \frac{\rho\, l}{A}$

where

| | |
|---|---|
| $R$ | is the resistance (Ω), |
| $\rho$ | is the resistivity (Ω m), |
| $l$ | is the conductor length (m) and |
| $A$ | is the cross-sectional area of the conductor (m$^2$). |

2. Transpose the formula if necessary. Not necessary in this case, as we have been asked for the resistance.

3. Convert data to SI units.

$\rho$ = 17.3 μΩ mm
= $17.3 \times 10^{-6} \times 10^{-3}$ Ω m ($10^{-6}$ to convert μΩ to Ω and $10^{-3}$ to convert mm to m)

$l$ = already in metres

$A$ = 2 $mm^2$
= $2 \times 10^{-6}$ $m^2$ ($10^{-3} \times 10^{-3}$)

Note that in the multiplication of the powers of 10 the two indices are added. In division they are subtracted.

4. Put in the data and calculate the answer.

$$R = \frac{17.3 \times 10^{-6} \times 10^{-3} \times 5}{1.5 \times 10^{-6}}$$

Note that the $10^{-6}$ on both top and bottom of the equation cancel out. (Dividing the top and the bottom of the equation by the same amount is called cancelling.)

$$= \frac{86.5 \times 10^{-3}}{1.5}$$

$$= 57.67 \times 10^{-3}$$

$$= 0.057\ 67\ \Omega$$

5. As the resistance has not been asked for in any specific unit we need not change it from the ohms we have calculated it in.

*Try this*

What is the resistance of a material with a resistivity of 17.8 μΩ mm, a cross-sectional area of 120 $mm^2$ and length of 100 m? Show your working.

## Calculators

A calculator is a useful resource tool.
It will perform number operations accurately.
*But* there are a few points to watch out for:

- The calculator cannot decide for itself which operation to use in any particular situation!
- If the data and/or the operation are entered incorrectly the calculator will happily accept them!
- The calculator will give you the answer, but it won't know if you require your answer to be adjusted in any way! For example, most answers can be rounded off to 2 significant decimal places (see below). Or you may have been required to give your answer in a multiple or submultiple of the unit calculated in.

So when you use your calculator, *remember*:

- Check out which operation you need to perform.
- Enter the data and operation correctly.
- Check that the answer is a sensible one for the question asked – estimate an approximate answer.
- If you have an answer containing more than the required decimal places then round up or down your answer to 2 significant decimal places.

## Significant decimal places

The following are examples of rounding off to 2 significant decimal places.

3.1799 would become 3.18
4.2431 would become 4.24
0.000 624 would become 0.000 62

The first rule is to note is that in a decimal fraction, such as 0.0056, the first figure after the decimal point that is not a zero is the first significant figure.

Then you should note that if the first of the figures to be discarded is 5 or greater then the previous figure is increased by 1.

62.785 would become 62.79

In 62.785 the 5 is to be discarded so the 8 is increased to 9.

3.996 would become 4.00

Where there have been parts of whole numbers used the zeros must be included in 2 significant places.

If your answer had been required to 4 significant places the same rules apply.

0.5261 76 would become 0.5262

These rules apply whether or not you are using a calculator to perform your number operations.

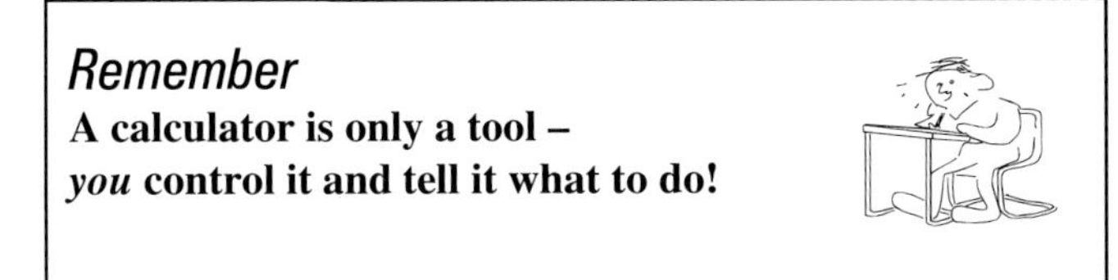

*Remember*

**A calculator is only a tool –**
***you* control it and tell it what to do!**

## The functions on a calculator

The +, –, × and = keys are very straightforward to use.

Try a few examples yourself. Remember to have an approximate answer and if appropriate round up or down any answers to 2 significant decimal places.

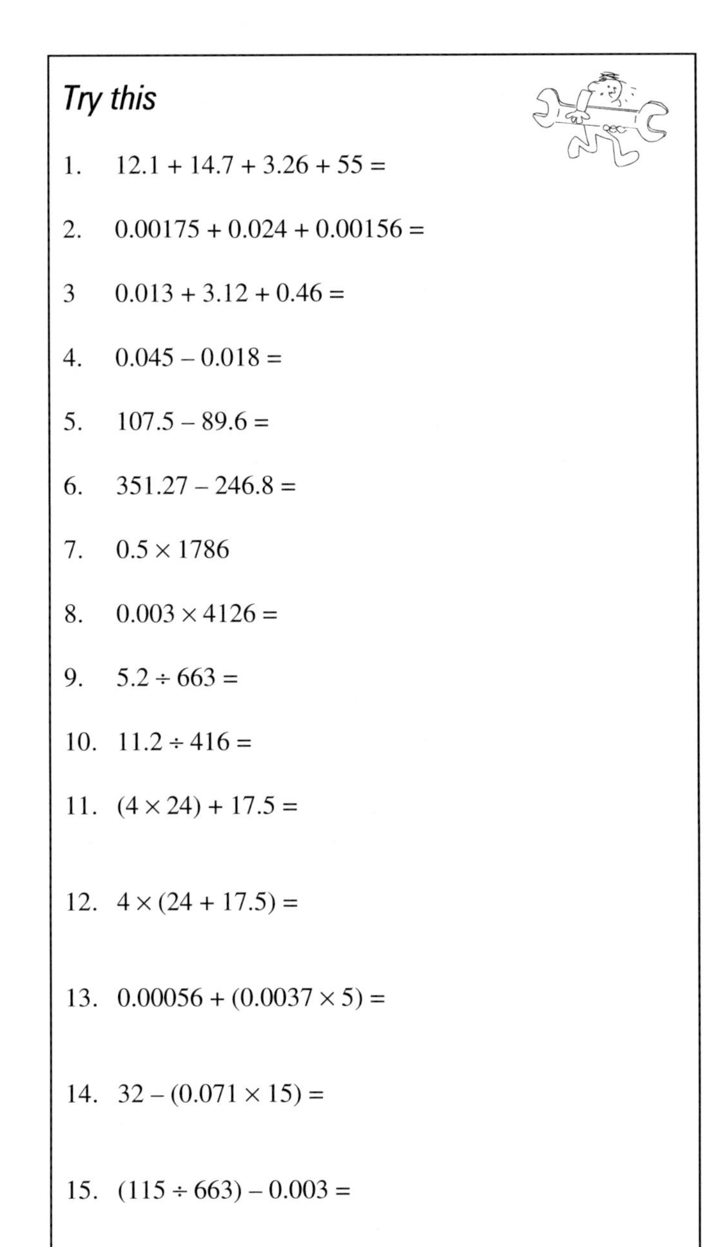

*Try this*

1. 12.1 + 14.7 + 3.26 + 55 =
2. 0.00175 + 0.024 + 0.00156 =
3. 0.013 + 3.12 + 0.46 =
4. 0.045 – 0.018 =
5. 107.5 – 89.6 =
6. 351.27 – 246.8 =
7. 0.5 × 1786
8. 0.003 × 4126 =
9. 5.2 ÷ 663 =
10. 11.2 ÷ 416 =
11. (4 × 24) + 17.5 =
12. 4 × (24 + 17.5) =
13. 0.00056 + (0.0037 × 5) =
14. 32 – (0.071 × 15) =
15. (115 ÷ 663) – 0.003 =

Try the following examples using your calculator.

### *Example (using square root key):*

The working resistance of a 1 kW fire element is 50 Ω. What current does this element take?

1. Write down the formula:

$$P = I^2 R$$

2. Transpose it if necessary.

$$P = I^2 R$$

$$I^2 = \frac{P}{R}$$

$$I = \sqrt{\frac{P}{R}}$$

3. Convert data if necessary (1 kW is 1000 W).
4. Substitute data and perform calculation. You need to evaluate what is under the square root sign before calculating the square root.

$$I = \sqrt{\frac{1000}{50}}$$

$$= \sqrt{20}$$

Using your calculator find the square root of 20.

The answer given by the calculator is

4.472 135 9

5. Adjust final value – round it off to 2 significant decimal places.

= 4.47 A

### Example (using percentage key):

The voltage drop in a circuit supplying a motor does not exceed 4%. Determine the maximum voltage drop in the case of a motor wired to a 230 V supply.

Maximum voltage drop = 4% of 230
(Enter 230 × 4%)
Maximum voltage drop = 9.2 volts

If you had been required to find the voltage at the motor terminals then you would have had to subtract the 9.2 V from the 230 V. This you could have combined with working out the percentage on the calculator.

Enter 230 – 4%.
Voltage at motor terminals = 220.8 volts

Adding on a percentage is just as easy. If you want to know how much your purchase will cost including 17.5% VAT when you are told it is £16.80 + VAT you enter 16.80 + 17.5 % and you find that the total cost is £19.74.

### *Example:*

The input power to drive a motor is 3500 W. If the percentage efficiency of the motor is 75%, what will the power output be?

Percentage efficiency = efficiency × 100
or transposing it

$$\text{efficiency} = \frac{\text{percentage efficiency}}{100}$$

$$= \frac{75}{100}$$

Formula required:

$$\text{efficiency} = \frac{\text{output}}{\text{input}}$$

Transpose formula

$$\text{output} = \text{efficiency} \times \text{input}$$

$$\text{output} = \frac{75 \times 3500}{100}$$

$$= 2625 \text{ W}$$

### *Try this*

**Using your calculator find answers to the following questions.**

1. Find the square root of 356.

2. What is 60% of 750?

3. Find the nominal diameter of copper wire of c.s.a. 120 $\text{mm}^{\mathbf{2}}$. Answer to be in mm.

4. The power output from a motor is 255 W and the input is 1.4 A at 230 V. Calculate the percentage efficiency of the motor.

5. In Figure 1.18 the power dissipated in resistor $R$ is 1000 W. Find the current.

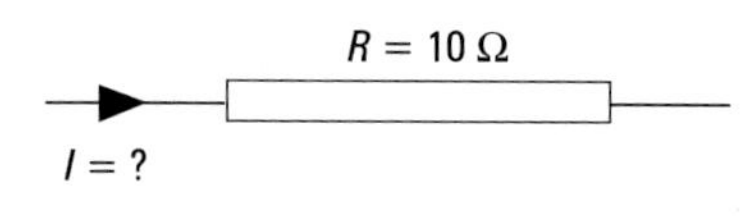


*Figure 1.18*

### *Self-assessment multi-choice questions*

**Circle the correct answers in the grid below.**

1. Speed is measured in the unit
   (a) metre/second$^2$
   (b) square metre
   (c) metre per second
   (d) cubic metre
2. The prefix “giga” is used to denote a multiple of
   (a) $10^{12}$
   (b) $10^9$
   (c) $10^6$
   (d) $10^3$
3. The prefix “micro” is used to denote a submultiple of
   (a) $10^{-12}$
   (b) $10^{-9}$
   (c) $10^{-6}$
   (d) $10^{-3}$
4. The cross-sectional area of trunking whose dimensions are 75 mm × 75 mm is
   (a) 5625 $\text{m}^2$
   (b) 56.25 $\text{m}^2$
   (c) 562.5 $\text{mm}^2$
   (d) 5625 $\text{mm}^2$
5. Transposing the formula
   $$P = I^2 R$$
   to find $R$ would result in
   (a) $R = \frac{P}{I^2}$
   (b) $R = \frac{I^2}{P}$
   (c) $R = PI^2$
   (d) $R = \sqrt{PI}$

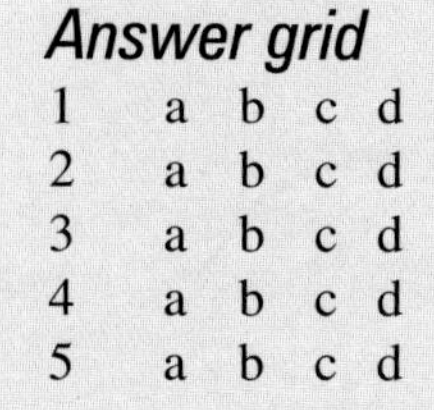

### *Answer grid*

| | | | | |
|---|---|---|---|---|
| 1 | a | b | c | d |
| 2 | a | b | c | d |
| 3 | a | b | c | d |
| 4 | a | b | c | d |
| 5 | a | b | c | d |

# 2

# Mechanical Science

Complete the following statements to remind yourself of some important facts from Chapter 1.

The metre is the SI unit for

The newton is the SI unit for

The farad is the SI unit for

The weber is the SI unit for

The tesla is the SI unit for

The ohm is the SI unit for

$10^6$ is ten to the power of six or 1 000 000

$10^3$ is ten to the power of or

$10^1$ is ten to the power of or

$10^{-9}$ is ten to the power of or

Note that all answers to questions in this book should be to 2 significant decimal places unless otherwise specified.

**On completion of this chapter you should be able to:**

- describe mass, force and weight
- state the effect of gravitational pull
- describe centre of gravity and stability
- complete exercises finding the mass of solids, liquids and gases
- define work and perform calculations
- describe the principle of levers and apply it to tools and equipment
- describe the principle of thermometers
- define and use in calculations the coefficient of linear expansion
- complete the revision exercise at the beginning of the next chapter

# Part 1

## Mass, force and weight

Mass is the quantity of matter that a body contains and it is measured in kilograms. The SI unit symbol is kg and the quantity symbol *m*. A body's mass can be found by weighing the body and comparing it with known standard masses (Figure 2.1).

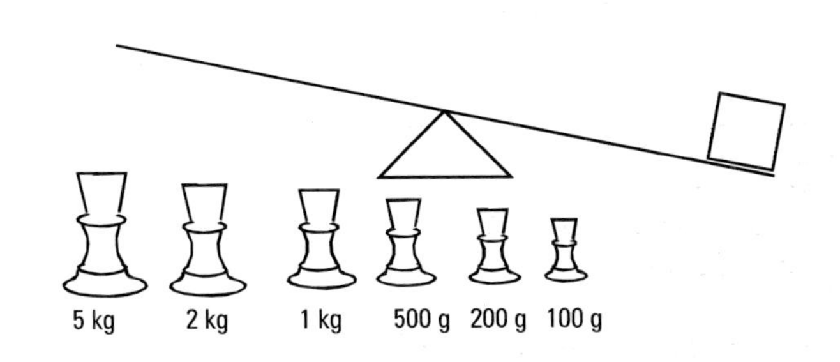


*Figure 2.1*

Force, at this stage, is best considered as an influence which tends to cause a body to move or change shape, and could be described as a push or pull. The SI unit of force is the newton, symbol N, named after the English scientist Isaac Newton.

If an object is stationary, or moving in the direction of an applied force the object will increase speed in the direction of the applied force (Figure 2.2). This is known as acceleration, and the greater the applied force the greater the acceleration.

*Figure 2.2*

If the object is already moving in the opposite direction to the applied force the effect of the force is to slow the body down, known as deceleration (Figure 2.3).

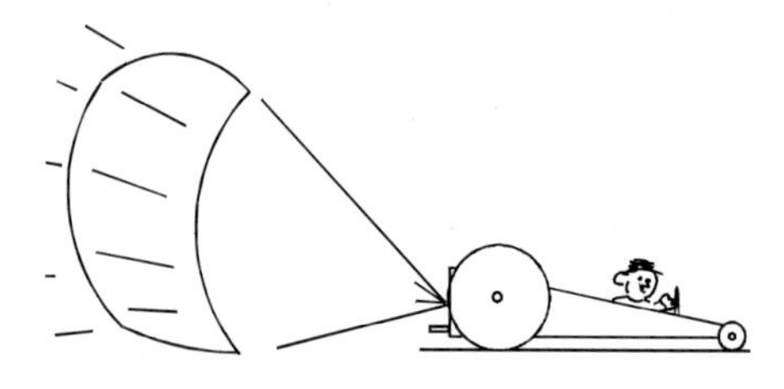

*Figure 2.3* *Parachute slowing Sid's dragster down*

If the body is unable to move then the applied force may cause distortion, as when bending a tube in a conduit bender (Figure 2.4).

*Figure 2.4* *Sid bending conduit*

If the object is unable to move or distort the applied force will be exactly resisted by other forces.

If either team in a Tug of War exerts more force than the other the rope marker and the opposing team will move; whilst the forces remain equal the teams remain static (Figure 2.5).

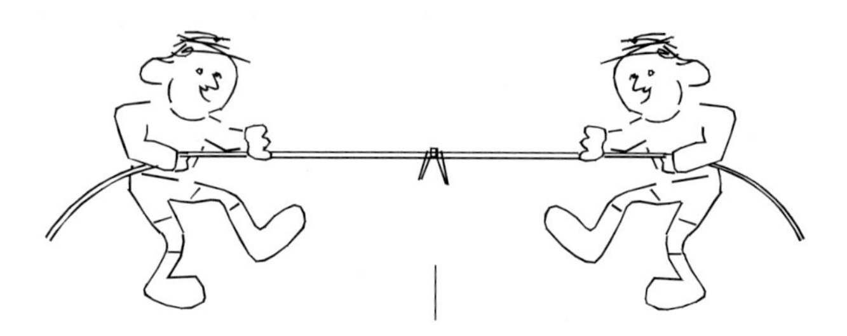

*Figure 2.5* *A Tug of War – teams are equal*

A brick left on level ground remains stationary. This is because the earth exerts a force on all masses. We know this force as gravity. The earth's "gravitational field" varies only slightly at any point on the earth's surface and it is generally accepted that a mass of 1 kg at the earth's surface experiences a force of 9.81 newtons due to gravity.

Weight can be considered as the product of the mass of an object and the force of the earth's gravitational force. In space, for example, it could be regarded as "weightless", although it would still have the same mass.

*Figure 2.6* *A brick placed in space stays where it is put*

In this situation (Figure 2.6) there would be no force acting upon the brick and it would remain precisely where we left it, even floating above a surface. If a force was applied to the brick it would move in the direction of the applied force (Figure 2.7). It would continue to move at a constant speed which would not change until a further force was applied.

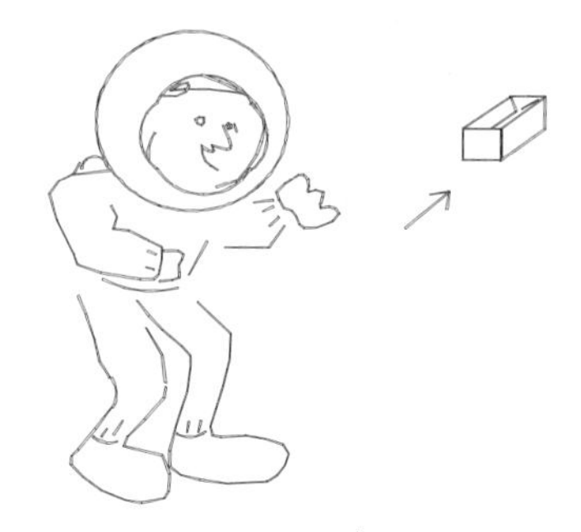

*Figure 2.7* *A slight push and it will keep going at the same rate until it meets a further force.*

**Inertia** is the name given to the property of an object continuing in an existing state of either rest or constant motion.

As we said earlier the force of an object may be considered as the product of the mass of the object and its acceleration. This can be expressed as

$$F = m \times a$$

where $F$ is the weight of the body in newtons and $m$ is the mass in kg and $a$ is the acceleration in $m/s^2$ (metres per second squared).

An object falling freely in the earth's gravitational field is taken to have an acceleration of $9.81\ m/s^2$, so

$$\text{newtons} = \text{mass in kilograms} \times 9.81\ m/s^2$$

**Pressure** The kilogram is the unit of mass and therefore cannot be used as a unit of force. The newton is the unit of force and we can consider that an object, with a mass in kg, will require a force in newtons to raise it against the force of gravity. It therefore follows that the object must exert this same force on any surface upon which it is placed.

The area of the object, along with the force exerted upon it exerts pressure on the surface upon which it is placed. This pressure is a product of the force and area and is therefore measured in $N/mm^2$.

### *Example*

A mass of 400 kg has a force of 50 N applied to it. What is the acceleration?

$$F = m \times a$$

Rearranging the formula

$$a = \frac{F}{m}$$

Substituting the known quantities

$$a = \frac{50}{400}$$
$$a = 0.125$$

The acceleration is 0.125 $m/s^2$.

### *Example*

A mass of 20 kg is given an acceleration of 4 $m/s^2$. What was the force required?

$$F = m \times a$$
$$F = 20 \times 4$$
$$= 80\ N$$

### *Example*

A motor with a mass of 120 kg is placed on a bench. Calculate the downward force on the table.

$$F = m \times a$$
$$F = 120 \times 9.81$$
$$= 1177.20\ N$$

### *Example*

Some trunking has a mass of 400 kg. What force will be required to lift it?

$$F = m \times a$$
$$F = 400 \times 9.81$$
$$= 3924.00\ N$$

### *Remember*

**MASS is the quantity of matter that a body contains.**
**FORCE is an influence tending to cause the motion of a body.**
**WEIGHT is the force experienced by a body due to the earth's gravitation.**

### *Try this*

1. A mass of 300 kg has a force of 30 N applied to it. What is the acceleration?

2. A mass of 80 kg is placed on a table. Calculate the downward force on the table.

3. What force would be required to lift a conduit having a mass of 350 kg?

A force has both magnitude and direction. We can show this as a line representing the force applied to a body. The length of the line, drawn to a suitable scale, represents the magnitude, or size, of the force and the direction of the force is shown by the direction of the line.

Figure 2.8 shows an example:

6 N

*Figure 2.8*

When using vectors to represent force on a body it is assumed that the force acts upon a single point, and the direction of the force is illustrated by the angle of the line with an arrow placed at the end of the line indicating the direction in which the force operates. If two forces act upon a body in the same direction the resultant will be the sum of the forces (Figure 2.9):

$$6\ N + 8\ N = 14\ N$$

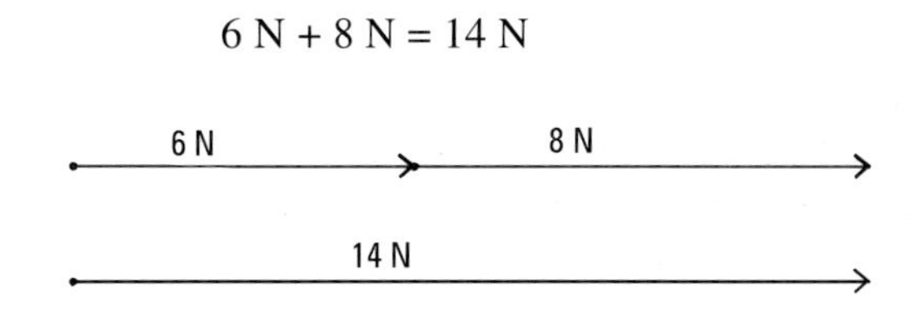


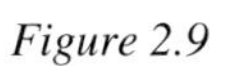

*Figure 2.9*

If these two forces act in opposite directions the result will again be the mathematical sum of the forces (Figure 2.10).

$$8\text{ N} - 6\text{ N} = 2\text{ N}$$

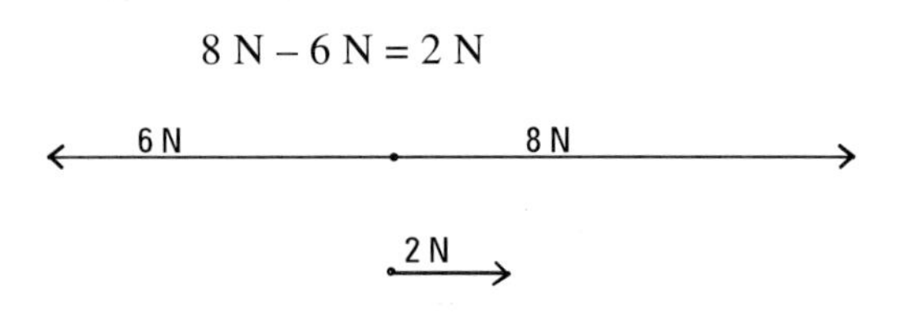


*Figure 2.10*

When the two forces are opposite and equal then the body is said to be in **equilibrium**, and there would be no resultant direction.

In reality, forces are often acting at an angle to one another. In such cases we often need to find the resultant force. We can do this by producing a parallelogram of the forces. Where two forces are operating at $90^{\circ}$ to one another, as in Figure 2.11, for example:

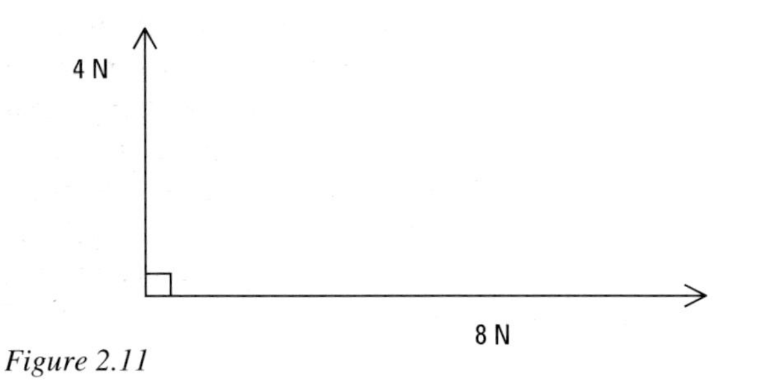


*Figure 2.11*

completing the parallelogram of forces we have Figure 2.12:

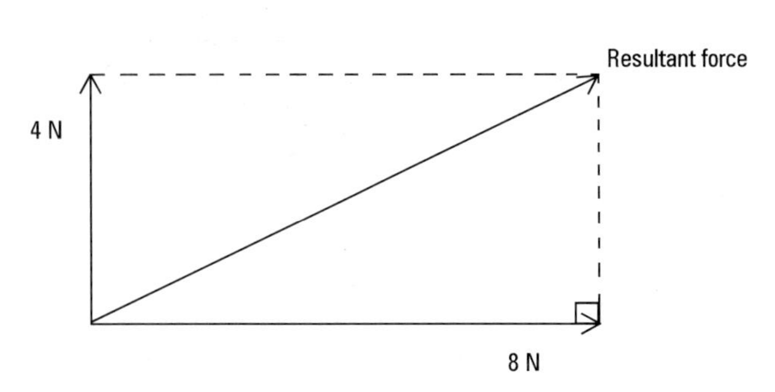

*Figure 2.12*

The resultant force is the diagonal of the parallelogram giving both the magnitude and the direction of the force. If the vector is drawn to scale the value of the resultant force may be measured, as may be the angle at which the force operates.

We can calculate the magnitude of the force using Pythagoras' theorem for right-angled triangles, where

$$\text{resultant force} = \sqrt{\text{force } 1^2 + \text{force } 2^2}$$

$$= \sqrt{8^2 + 4^2}$$
$$= \sqrt{64 + 16}$$
$$= \sqrt{80}$$
$$= 8.94\text{ N}$$

We can apply the parallelogram of forces irrespective of the angle at which the forces operate providing we remember the basic rules of keeping the angle and length the same for the parallel sides. For example (Figure 2.13):

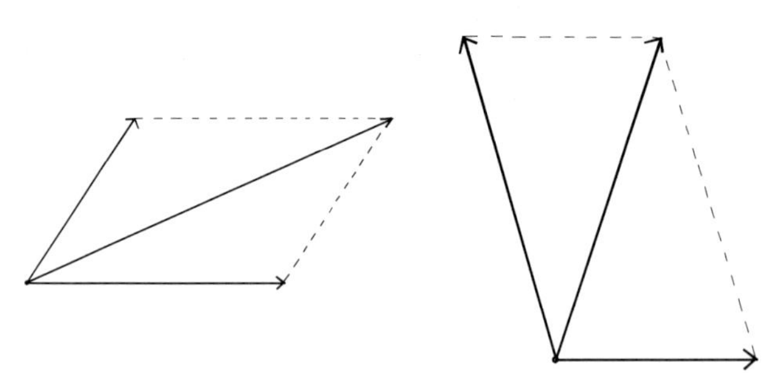

*Figure 2.13*

## *Try this*

Use a scale of 1 cm representing 1 N and complete the parallelogram for the following forces. Draw in the diagonal and measure the resultant force.

1. Forces of 4 N and 3 N at an angle of $60^{\circ}$ to each other.

2. Forces of 2.5 N and 5 N at an angle of $90^{\circ}$ to each other.

Let us consider a practical application of this information.

A pole supports an overhead cable at a point where the cable changes direction through 90°. A stay wire is needed to ensure that the pole is not pulled over by the force of the cables (Figure 2.14). First we need to know what force is going to act upon the stay wire and at what angle it must be fixed to keep the pole upright.

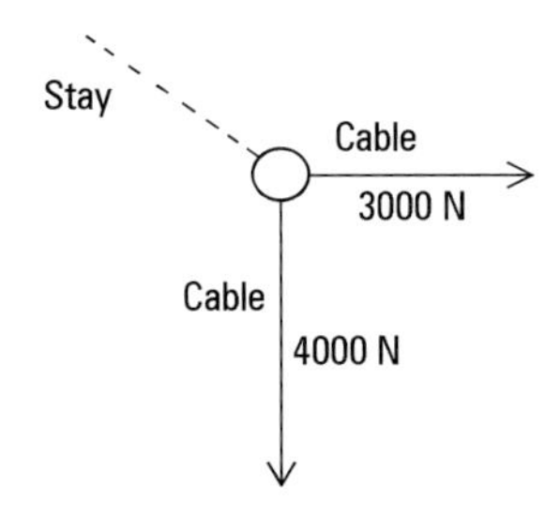


*Figure 2.14*

If the forces from the cable are 3000 N and 4000 N, then to calculate the force and direction of the pull on the pole we can use our parallelogram of forces (Figure 2.15).

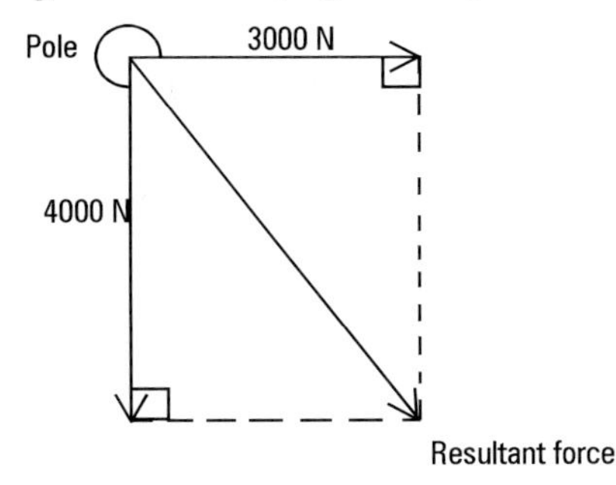


*Figure 2.15*

Resultant force $= \sqrt{4000^2 + 3000^2}$

$= 5000$ N

If the stay wire is fixed in opposition to the resultant force at an angle of 45° (Figure 2.16), what is the force exerted on the anchor?

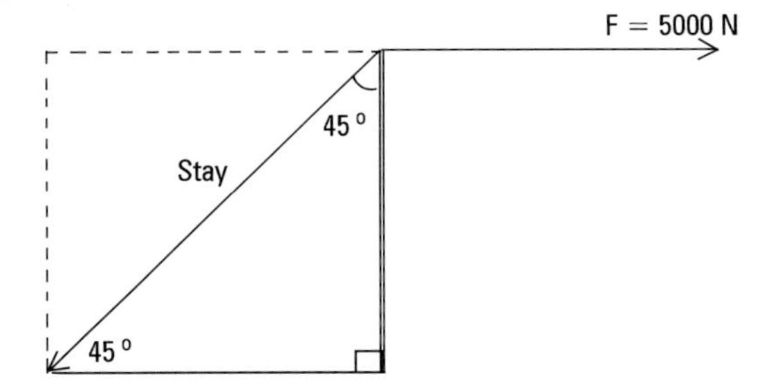


*Figure 2.16*

Tension force in stay $= \sqrt{5000^2 + 5000^2}$
$= \sqrt{50\,000\,000}$
$= 7071$ N

**Stress** The forces applied to the stay wire in our example have a tendency to stretch the material, as the forces are in opposition and away from the material of the wire (Figure 2.17).

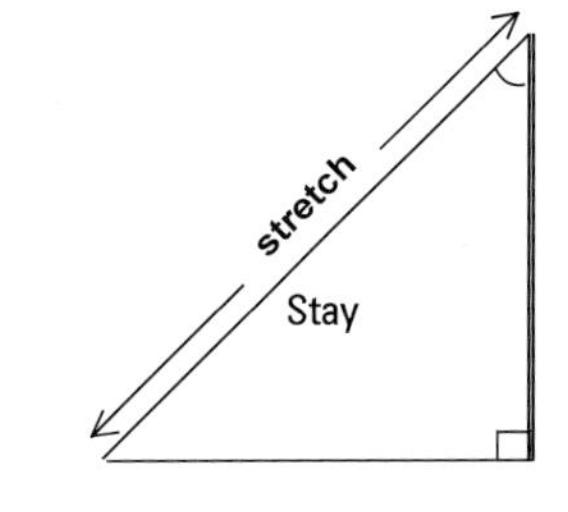


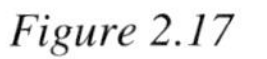

*Figure 2.17*

This places the wire under stress and where the stress tends to stretch the material we refer to this as "**tensile stress**". If the force applied tends to compress the material, like squashing it in a vice, then we refer to this as "**compressive stress**".

There is one further application of force which produces stress which we need to consider. This is called **shear stress**. When we use tin snips to cut metal capping we are applying shear stress to the metal, and in this case the force applied is greater than the material can withstand. Another example of shear stress is the forces experienced by a bolt or rivet joining materials together. In our stay wire example, bolts holding together the link arrangement are placed under shear stress (Figure 2.18).

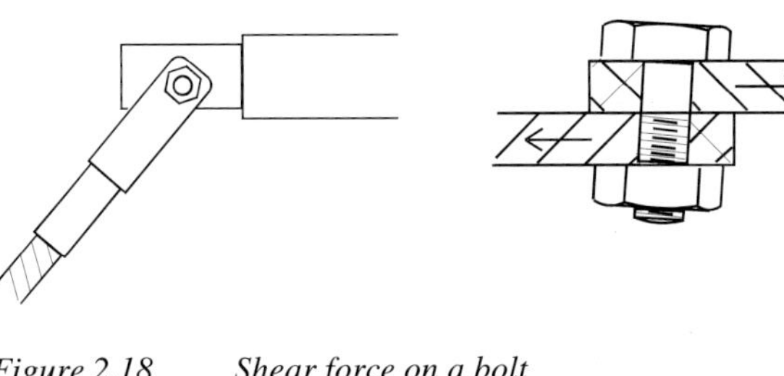

*Figure 2.18* *Shear force on a bolt*

## *Points to remember*

Define the following:

Mass is

Force is

Weight is

Force is a vector quantity. This means it has both magnitude and direction.

# Part 2

## Density

Density is the measure of how compact a substance is. If we compare two blocks of the same size, one made of iron and the other wood, the iron block would be the heaviest, so we can say that iron has a greater density than wood.

The SI unit for density is kg/m$^3$ (kilograms per cubic metre). Its symbol is ρ (rho) and the sub-multiples are kg/dm$^3$ (kilograms per cubic decimetre) and g/cm$^3$ (grams per cubic centimetre).

$$\text{density} = \frac{\text{mass}}{\text{volume}}$$

$$\rho = \frac{m}{V}$$

Follow through these examples before answering a question by yourself.

### *Example*

If the density of copper is 8900 kg/m$^3$, what will be the mass of a block of copper 0.2 m × 0.2 m × 0.1 m?

$$\text{density} = \frac{\text{mass}}{\text{volume}}$$

First we must find the volume of the block.

$$0.2 \times 0.2 \times 0.1 = 0.004\,\text{m}^3$$

Now rearranging the formula:

$$\text{mass} = \text{density} \times \text{volume}$$

Fill in the known quantities:

$$\text{mass} = 8900 \times 0.004$$
$$= 35.6\ \text{kg}$$

### *Example*

If the density of air is 1.3 kg/m$^3$ what is the mass of air in a room 3 m × 4 m × 15 m?

$$\text{Volume of the room} = 3 \times 4 \times 15$$
$$= 180\ \text{m}^3$$

$$\text{mass} = \text{density} \times \text{volume}$$
$$m = 1.3 \times 180$$
$$= 234\ \text{kg}$$

### *Example*

A water tank has dimensions of 3 m × 2 m × 4 m. What is the maximum quantity of water (in litres) it can hold?
Water has a density of 1000 kg/m$^3$ and 1 litre of water has a mass of 1 kg.

$$\text{volume of tank} = 3 \times 2 \times 4$$
$$= 24\ \text{m}^3$$

$$\text{mass} = \text{density} \times \text{volume}$$
$$= 1000 \times 24$$
$$= 24\,000$$
$$\text{mass} = 24\,000\ \text{kg}$$

As 1 litre of water has a mass of 1 kg, the tank will hold a maximum of 24 000 litres of water.

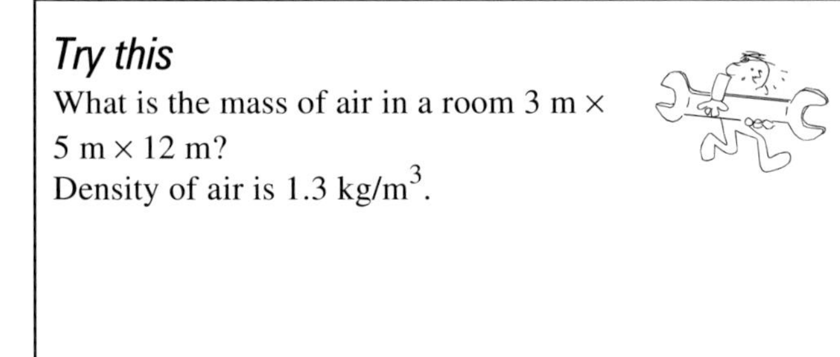

### *Try this*

What is the mass of air in a room 3 m × 5 m × 12 m?
Density of air is 1.3 kg/m$^3$.

## Levers

Levers, in the form of crowbars, are used in order to raise a heavy load with a small effort. The same principle is used by clawhammers (Figure 2.19).

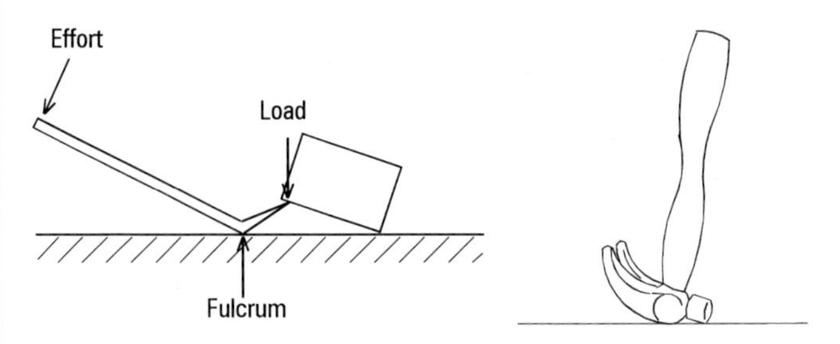


*Figure 2.19*

A lever is a bar pivoted so as to be able to rotate about a point. This point is called the fulcrum (Figure 2.20).

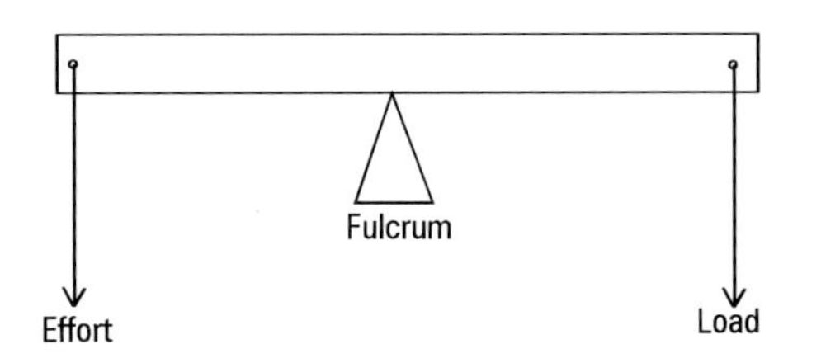


*Figure 2.20*

The position of the fulcrum is important. The nearer it is to the load the less force has to be applied in order to lift the load.

There are three common arrangements for simple levers. The crowbar is a first-order (or class 1) lever and the fulcrum is between the load and the effort. Second-order levers (Figure 2.21) are where the fulcrum is at one end and the load is nearest to the fulcrum. A wheelbarrow is an example of a second-order lever. In a third-order lever arrangement (Figure 2.22) the load is at the opposite end of the lever to the fulcrum, and in this case the effort is always greater than the load. Tweezers or woolshears are examples of third-order levers.

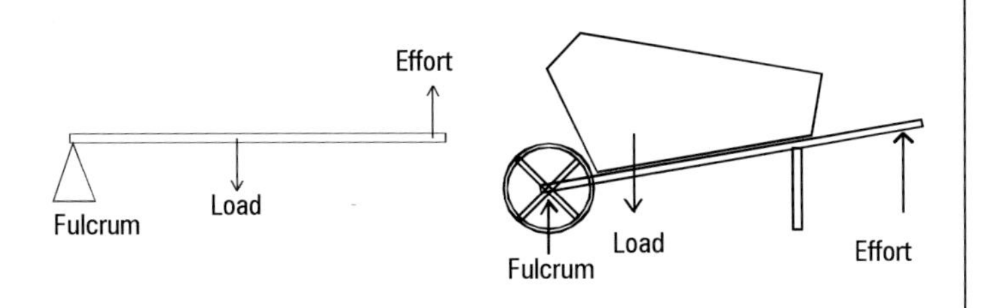


*Figure 2.21* *Second-order levers*

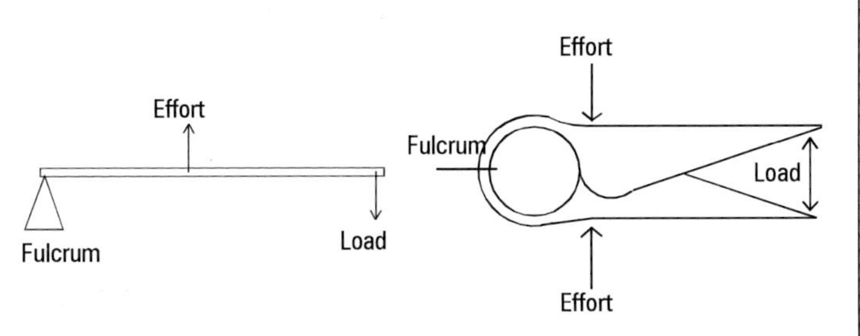


*Figure 2.22* *Third-order levers*

## Moments

The turning effect of the force (Figure 2.23) is called the moment of the force. A moment is measured in newton metres (N m).

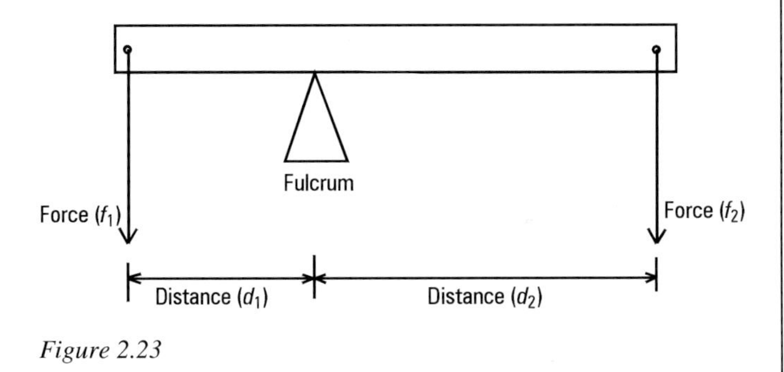


*Figure 2.23*

$$\text{moment} = \text{force} \times \text{distance}$$

$$= f_1 d_1 \quad \text{or} \quad f_2 d_2$$

A moment can cause a clockwise or an anticlockwise turn around the pivot or fulcrum. An anticlockwise moment is shown in Figure 2.24.

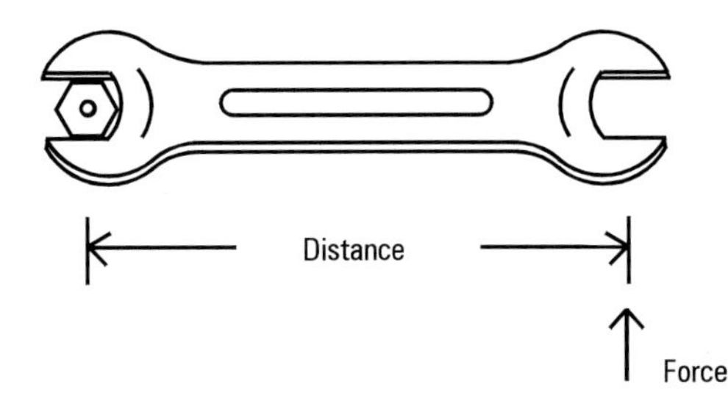


*Figure 2.24*

A state of **equilibrium** is said to exist if the anticlockwise moment equals the clockwise moment. This is called the **principle of moments**. In Figure 2.25 the downward forces are $F_1$ and $F_2$. The distances between the fulcrum and the downward forces are $d_1$ and $d_2$. In a state of equilibrium:

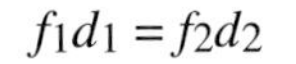

$$f_1 d_1 = f_2 d_2$$

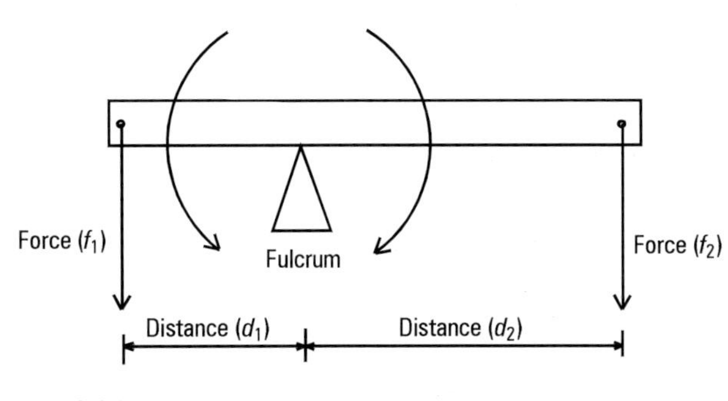


*Figure 2.25*

### *Example*

What effort would be needed to lift a load of 25 kg placed 200 mm from the fulcrum of the lever? The distance between the fulcrum and the point the force is to be exerted is 1 metre. Remember the force exerted by the 25 kg load is $25 \times 9.81$ newtons.

$$f_1 d_1 = f_2 d_2$$

$$f_1 = \frac{f_2 d_2}{d_1}$$

$$= \frac{25 \times 9.81 \times 0.2}{1}$$

$$= 49.05 \text{ N}$$

## *Try this*

What effort would be needed to lift a load of 25 kg placed 500 mm from the fulcrum of a lever? The distance between the fulcrum and the point the force is to be exerted is 900 mm.

## Centre of gravity

Every object has a centre of gravity, but this will not always be at its centre. The centre of gravity is determined by the shape of the object and its mass at different points.

Look at the diagrams below (Figure 2.26). Which ones seem **stable** and which ones **unstable**? The stable ones are those with a wide base and where the centre of gravity is low down.

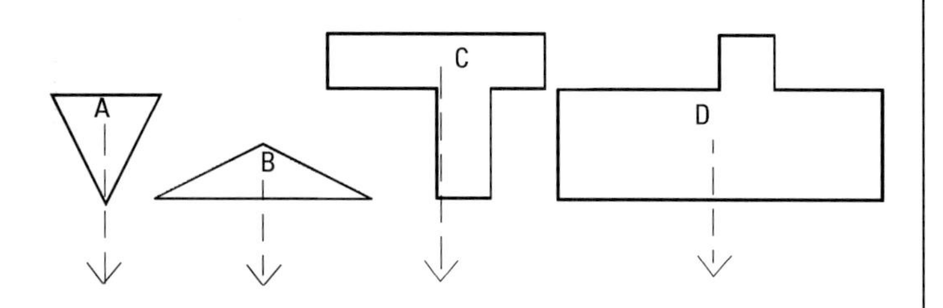


*Figure 2.26*

A and C will easily overbalance, as their centres of gravity need to be moved only slightly and they will no longer be directly over the point of contact with the surface.

For stability, a large area of the object needs to be in contact with the surface on which it is standing and the object must have a low centre of gravity. The centre of gravity needs to be taken into account when lifting or moving loads either manually or by assisted moving and when stacking objects (Figure 2.27).

*Figure 2.27 Take care when adding to this stack!*

## Work

The movement of a body by a force is described as work. The amount of work done is directly proportional to the force applied and the distance moved. The unit of work is the joule, unit symbol J.

Work done (joules) = force (newtons) × distance (metres)

$$W = Fd$$

## *Example*

What is the work done when a motor is pulled along the floor a distance of 2 m by a force of 20 N?

$$W = Fd$$
$$= 20 \times 2$$
$$= 40 \text{ joules}$$

## *Remember*

- A moment is the turning effect on an object produced by a force acting at a distance.
- A lever is a rigid bar pivoted about a fulcrum which can be acted upon by a force in order to move a load.
- A fulcrum is the point against which a lever is placed.
- The centre of gravity is the point in a body about which the mass is evenly distributed.

## Energy

The capacity to do work is called energy, which comes in many forms. Some of the most significant forms of energy are atomic energy, chemical energy, heat energy, mechanical energy and electrical energy. To use energy we often need to convert energy from one form to another. Take, for example, a coal-fired power station (Figure 2.28).

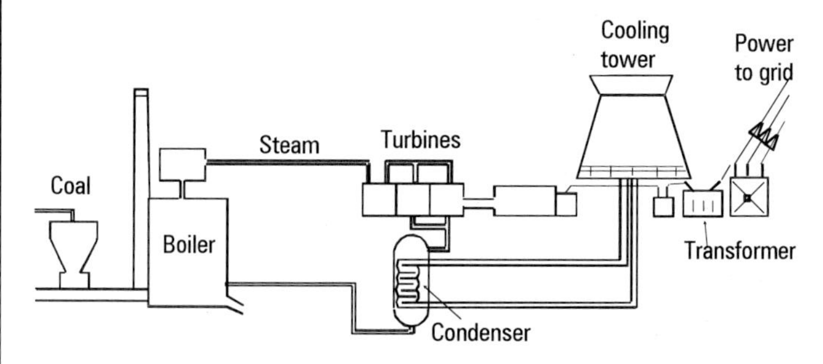


*Figure 2.28 Coal-fired power station*

Coal contains chemical energy which is converted, through combustion, into heat energy. This heat energy is used to produce steam, which in turn is the energy used to drive a steam turbine which converts the heat energy into mechanical energy. The turbine drives a generator which converts the mechanical energy into electrical energy.

During this conversion, energy can be neither created nor destroyed, so the total energy output is equal to the energy input. In Chapter 1 we saw that the useful energy output of a machine is less than the input. This is due to the losses within the machine. These losses are generally in the form of heat energy (friction) and noise energy (sound).

## Power

Another important factor when carrying out work is how quickly the work is carried out. Power is the measure of the rate at which work is carried out and is the result of the amount of work done and the time taken to carry out the work.

$$\text{power} = \frac{\text{work done}}{\text{time taken}}$$

The SI unit of power is the watt (symbol W), and one watt equals one joule per second or one newton metre per second.

### *Example*

The force required to raise a load through a vertical distance of 8 m is 50 N. The time taken for this is 40 s. Calculate the power required.

$$\text{power} = \frac{\text{work done}}{\text{time taken}}$$

$$\text{work done} = 50 \times 8 = 400\text{ J}$$

$$\text{power} = \frac{400}{40} = 10\text{ W}$$

### *Points to remember*

**Density**

Complete the formula:

mass = density × __________

**Levers**

A lever is a bar pivoted so as to be able to pivot about a point. This point is called the ______________

**Centre of gravity**

The centre of gravity of an object is determined by the shape of the object and its mass at different points. Draw an object which appears stable and one which appears unstable.

**Energy**

Name as many forms of energy that were mentioned in this chapter as you can remember.

### *Try this*

1. What is the work done when a load of 8 kg is lifted 1.5 metres from the ground?
   (Remember:
   weight in newtons = mass in kilograms × 9.81 $m/s^2$.)

2. What is the power output of a hoist that can lift a mass of 20 kg a height of 15 m in 40 s?

3. The cage of a lift weighs 1000 kg. The maximum load it can carry is 2000 kg. If the lift is carrying a load of one quarter its maximum, what is the work done if the lift is raised to a height of 30 metres? If this is achieved in a time of 40 seconds what was the power required?

# Part 3

## Temperature

Temperature is the measure of heat in a body and can travel between bodies in three ways;

**conduction**

placing one end of a poker in an open fire for a period of time results in the handle becoming warm by conduction through the material

**radiation**

if you stand in front of the open fire the "heat" from the fire can be felt radiating from the fire

**convection**

a central heating radiator raises the air temperature within a room partly by convection, warming the air, causing it to rise and create air circulation within the room

When heat energy is added to a body it produces a change in the body. This may be a change in temperature or a change in the state of the body. For example, if heat is applied to ice that is below 0 °C the temperature of the ice will rise until it reaches 0 °C. At that point no further rise in temperature will occur until all the ice has melted. When all the ice has melted the temperature of the water will continue to rise until it reaches 100 °C when the water turns to steam. The temperature then remains at 100 °C until all the water has been turned into steam.

*Remember*

**HEAT is a form of energy.**
**TEMPERATURE is measured by use of a thermometer and is an assessment of how hot a substance is.**

## Change due to pressure

It is also worth remembering that pressure can produce a change in state. When gas is pressurised it can reduce in volume and become a liquid. This effect is often used for transportation and storage, as in fire extinguishers and oxygen cylinders.

## Change in volume due to temperature

Generally, when heat is applied to solids, liquids and gases they expand and increase in volume. This effect is put to good use in thermometers and temperature control devices. A thermometer may use a glass tube with a bulb of mercury. When the thermometer is heated the mercury expands and the level in the tube rises, indicating the temperature. Alcohol may be used in place of mercury where lower temperature measurements are required.

The SI unit of temperature is the Kelvin, but as we found earlier it is common to use the degree Celsius.

Remember: 0 °C = 273.15 °K

Although there are some special materials which react differently the majority of materials expand when heated. If we take a length of a material and raise its temperature, the length of the material increases. The amount of increase is dependent upon the temperature rise and the type of material. The expansion of a given material at a given rate is dependent upon its composition. The effect is called the **coefficient of linear expansion** and is "the expansion per unit length of material for unit temperature rise" (symbol $\alpha$ – alpha).

$$\alpha = \frac{\text{change of length}}{\text{original length} \times \text{temperature rise}}$$

Different materials have different coefficients of linear expansion.

For example

iron $\alpha$ is approximately $11.2 \times 10^{-6}$ per °C in the temperature region of 20 °C

copper $\alpha$ is approximately $17 \times 10^{-6}$ per °C.

The principle that different metals expand at different rates is used for many temperature control devices and thermostats.

### *Example*

A copper busbar 2 m long increases in length (Figure 2.29) when it is heated through 100 °C by an electric current. If the coefficient of linear expansion of copper is $17 \times 10^{-6}$/°C, what was the increase in length?

Using the formula

$$\alpha = \frac{\text{change of length}}{\text{original length} \times \text{temperature rise}}$$

and transposing it

$$\text{change of length} = \alpha \times \text{original length} \times \text{temperature rise}$$

$$= 0.000017 \times 2 \times 100$$
$$= 0.0034 \text{ m or } (3.4 \text{ mm})$$

Expansion

*Figure 2.29*

## *Try this*

At 10 °C a bar of mild steel is 500 mm long. At what temperature will its length be increased to 500.2 mm? The coefficient of linear expansion of mild steel is $11 \times 10^{-6}/°C$.

## Temperature control devices

Heating appliances, such as cookers and water heaters, generally use temperature control devices which rely upon expansion. The material used in the device may be gas, liquid or metal, dependent upon the application of the control device. Some of these controls, such as thermostats, operate on actual temperature, while others, such as simmerstats, operate on comparative temperature.

An oven thermostat (Figure 2.30) may have a bulb filled with gas or liquid which is placed within the oven. This is connected to the control switch by a capillary tube, expansion causes the control bellows to open and close, operating the switch controlling the heat. A similar arrangement is used to control a refrigerator or freezer, except that the switch operation is set to switch on when temperature rises and off when the temperature falls.

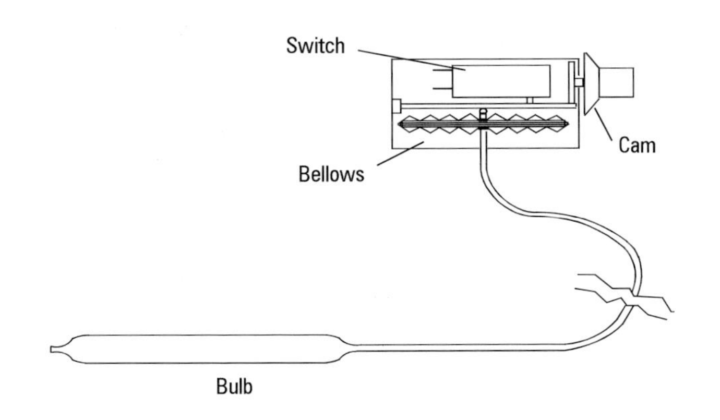


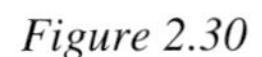

*Figure 2.30*

Room thermostats (Figure 2.31) used to control central heating systems commonly use a bimetal switch to operate the control switch. This relies on the different expansion rates of two different materials, fixed together. When heated the result is a bending of the strip operating the switch.

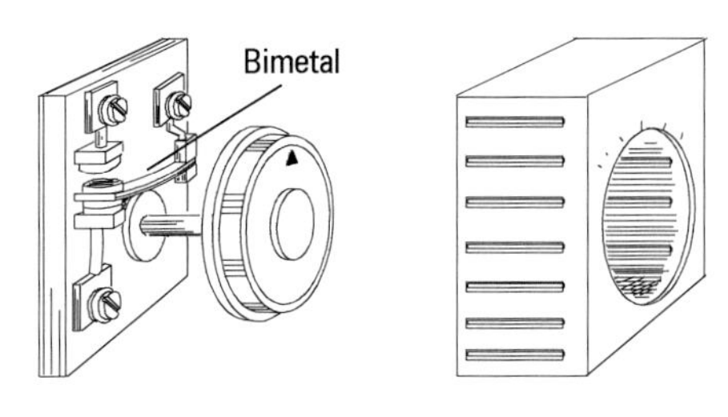

*Figure 2.31 Room thermostat*

The basic immersion heater thermostat also relies on the different expansion rate of materials. They are often referred to as rod thermostats (Figure 2.32), as they operate by use of a nickel alloy rod within a brass tube. The shape of the thermostat allows it to be placed inside a tubular pocket within the water. The brass tube and alloy rod are joined at the end furthest from the switch. The outer brass tube expands at a higher rate than the inner alloy rod. The result is that the rod is drawn away from the contacts, causing them to open.

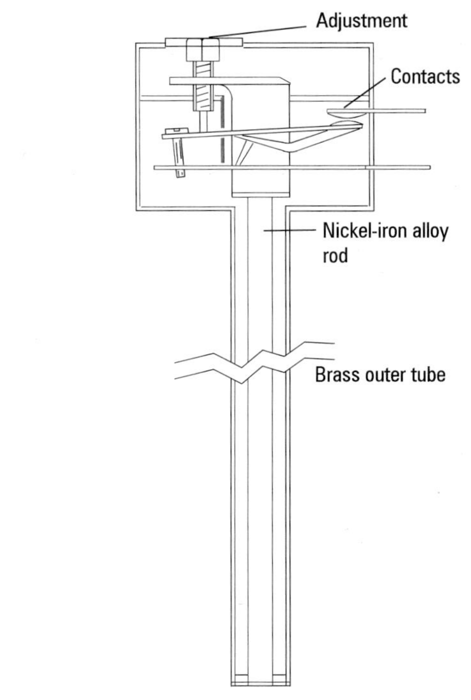

*Figure 2.32 Rod thermostat*

The simmerstat is different from the thermostat because, while the thermostat controls heat, the simmerstat controls energy. A simmerstat can continue to work even when no load is connected. This is because the simmerstat contains its own source of heat, generally a resistance wire, which controls the action of a bimetal strip.

A simmerstat (Figure 2.33) is used to limit the heat of a cooker hot-plate. The control can be given a number of markings to show how high or low the setting is.

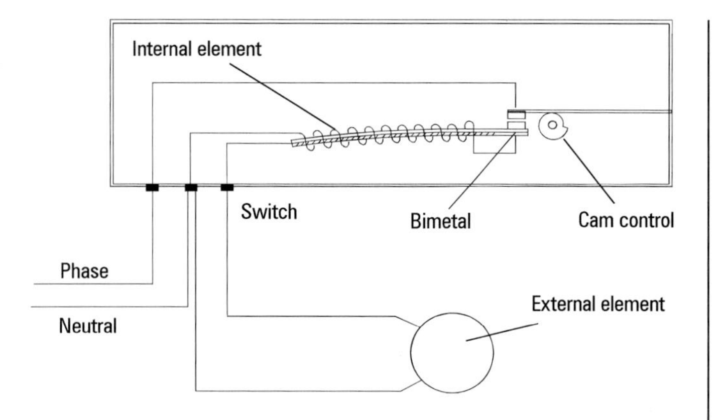


*Figure 2.33 Simmerstat*

## *Try this*

Explain in your own words the operation of a simmerstat.

## *Points to remember*

Complete the following:

## Mass, force and weight

Mass is the
and is measured in
Force is measured in
Weight is
and is measured in
The property of a body continuing in an existing state of rest or uniform motion is known as
If forces on a body are equal and opposite the body is said to be in a state of

## Shear force

Give two examples of shear force.

## Density

Density is the
and is measured in

## Levers and moments

A lever is
A fulcrum is
A moment is measured in

## Work and power

The work done is the force applied to an object multiplied by

Power is
The SI unit of power is the

## *Self-assessment multi-choice questions*

**Circle the correct answers in the grid below.**

1. A bundle of conduit has a mass of 200 kg. What force will be required to lift it?
   (a) 200 N
   (b) 20.39 N
   (c) 1962.00 N
   (d) 1962 kg
2. A mass of 1 kilogram experiences a force due to gravity of
   (a) 981 kg
   (b) 9.81 kg
   (c) 981 N
   (d) 9.81 N
3. A mass of 400 kg has a force of 20 N applied to it. What is the acceleration?
   (a) 800 $m/s^2$
   (b) 200 $m/s^2$
   (c) 20 $m/s^2$
   (d) 0.05 $m/s^2$
4. The SI unit for density is $kg/m^3$ and its symbol is
   (a) $\rho$
   (b) $\pi$
   (c) $\alpha$
   (d) $\eta$
5. Which of the following is affected by the output temperature of the device it controls?
   (a) simmerstat
   (b) circuit breaker
   (c) three-heat switch
   (d) thermostat

### *Answer grid*

| | | | | |
|---|---|---|---|---|
| 1 | a | b | c | d |
| 2 | a | b | c | d |
| 3 | a | b | c | d |
| 4 | a | b | c | d |
| 5 | a | b | c | d |

# 3

# Electrical Science

Answer the following questions to remind yourself of what was covered in Chapter 2.

1. What is the downward force on a bench of a motor with a mass of 150 kg?

2. If the density of copper is 8900 kg/m$^3$, what will be the mass of a block of copper 20 cm × 10 cm × 10 cm?

3. A bar 1 m long is pivoted at its centre. A downward force of 80 N is applied at right angles 0.2 m from one end. Calculate the downward force to be applied at right angles to the bar at the opposite end to prevent it from rotating. Ignore the weight of the bar.

4. A force of 0.2 N is used to move an object through 0.1 m in 4.5 s. Calculate the work done and the power.

**On completion of this chapter you should be able to:**

- relate electron flow to current flow
- distinguish between conductors and insulators
- describe the essential features of an electrical circuit
- identify the effects of current flow
- recognise the correct connections of voltmeters and ammeters
- calculate the factors related to Ohm's Law
- state the relevant units in SI terms
- complete the revision exercise at the beginning of the next chapter

## Part 1

Have you ever considered why some materials are used to conduct electricity and others act as insulators?

Some things, like water, can be used as either, depending on how pure they are.

Everything we use is made up of molecules which contain atoms. It is the structure of these atoms that can give us the answers to the above question.

Water is a good material to start with, for it contains two types of atoms, hydrogen and oxygen. The chemical formula for water is $H_2O$. This means it has twice as many hydrogen atoms as oxygen.

Let's first consider hydrogen.

Hydrogen has the atomic number 1. This means it has one set of components making it up. All atoms consist of two main component parts: the central nucleus, which contains the protons and neutrons, and the orbiting satellites, called electrons. The hydrogen atom has one proton and one electron. The electron revolves around the nucleus in a three-dimensional orbit (Figure 3.1).

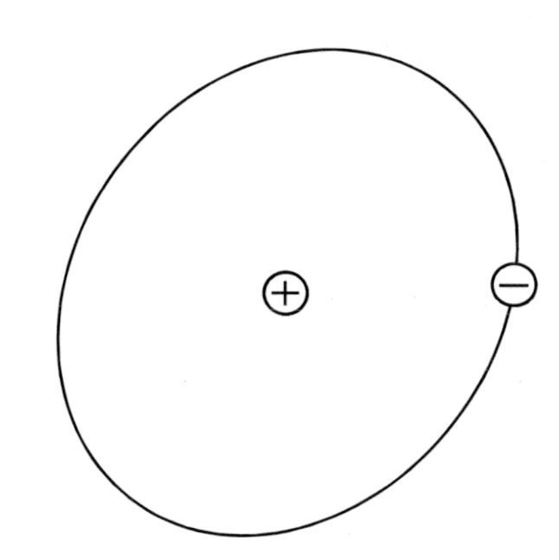

*Figure 3.1*

As protons are positively charged, neutrons are neutral (neither positive or negative) and electrons negatively charged, the atom overall has a neutral charge. All atoms in their neutral state have the same number of protons as electrons. The hydrogen atom is by far the simplest to follow with just one proton and one electron.

Now let's look at an atom of a material that we know is a good electrical conductor, copper.

The copper atom, with an atomic number of 29, is far more complex than the hydrogen atom. Its nucleus consists of 29 protons (positive charges), and when it is in its neutral state the copper atom has 29 electrons. The electrons are spread out across four orbits, or shells, as shown in Figure 3.2.

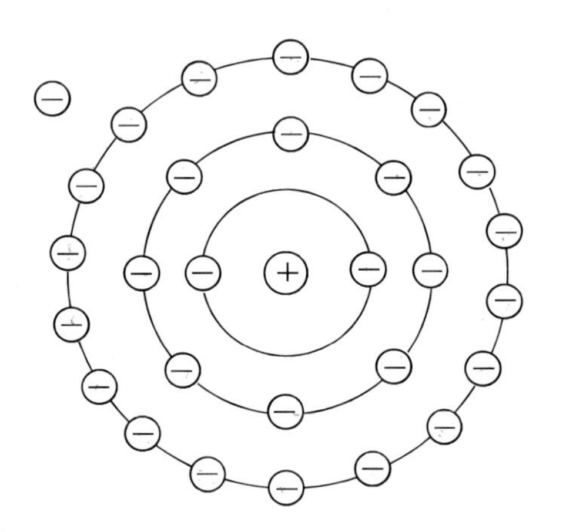

*Figure 3.2*

In the outermost orbit there is only one electron, and this is shielded from the nucleus by the other electrons. The effect of this is to make the outer electron very loose in its shell and easily affected by other atoms. Remember that the electrons are on three-dimensional orbits and each orbit is an ellipse. This means that the outer electron is often closer to the nucleus of another atom than its own and so is attracted to the outer shell of that atom.

So, in copper, the outer electrons of the atoms move around from one atom to another. These are called free electrons. When an electron, which is negatively charged, leaves an atom, the 29 positively charged protons are stronger than the remaining electrons, so the atom becomes positively charged. This is referred to as a positive ion. In this state the atom is very attractive to negatively charged free electrons from other atoms.

When copper is not being used as an electrical conductor the free electrons move about in a completely free fashion in any direction throughout the metal.

## Electron and current flow

The free negatively charged electrons, like all negative electrical charges, are attracted by positive charges.

An electric battery cell is basically a chemical unit which has two terminals (Figure 3.3). One of these has a surplus of electrons, the other a surplus of positively charged ions. When a copper conductor is connected across a cell the free electrons in the conductor are attracted to the positive terminal of the cell. This creates an electron flow in the conductor from negative to positive.

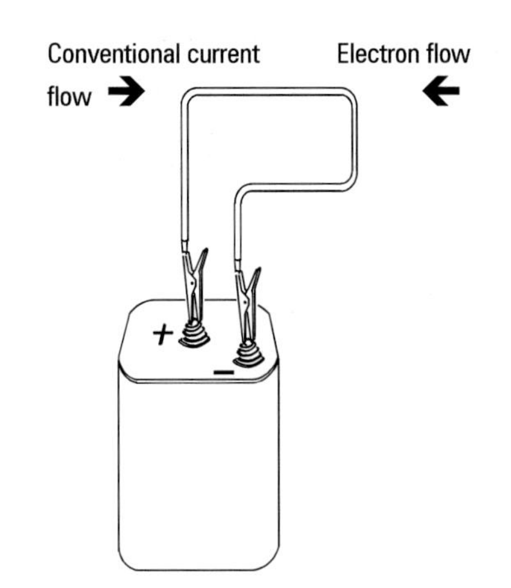


*Figure 3.3*

Although we now know that electron flow and current flow amount to the same thing, convention accepts that current flow is from positive to negative.

This is confusing, but remember that electrons are negative and are attracted to the positive. Current we consider is the opposite.

**Safety**
**Never short out a battery or cell.**

## The circuit

In practice we never short out a battery or cell with a conductor like copper. We need to have some kind of load connected, such as a lamp (Figure 3.4).

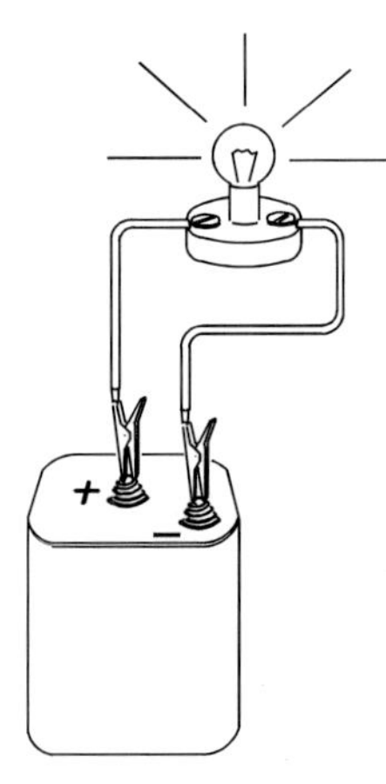


*Figure 3.4*

All the time the circuit is complete the electrons flow and the lamp lights. If any part of the circuit becomes disconnected it is "opened" and the electrons no longer flow in a single direction; current ceases to flow and the lamp goes out.

Here we talk about a circuit being a continuous length of conducting material which starts and ends at the different poles of a source of supply. The conductor may include the filament of a lamp, as in Figure 3.4, or the windings of a motor, as in Figure 3.5.

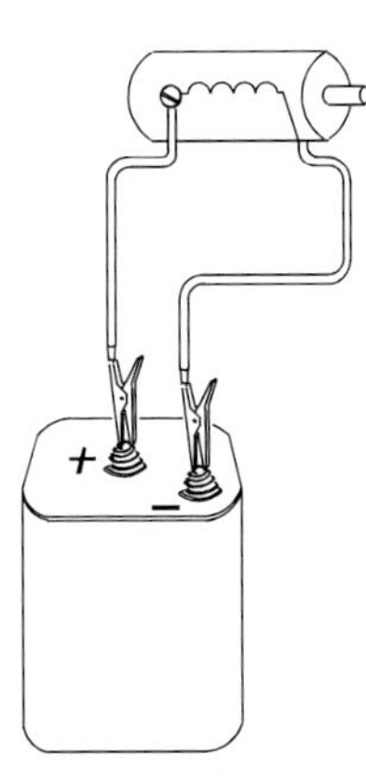


*Figure 3.5*

## Control of circuits

In addition to the conductor there is normally some type of control device. The simplest form of control is a switch. Switches are basically conductors that make and break the flow of electrons. As we have already seen, when a circuit is opened electrons cease to flow in a single direction, so current flow stops. A switch opens the circuit to stop current flow and closes the circuit to allow current to flow.

The circuit diagram (Figure 3.6) shows how the basic components of a simple circuit are connected.

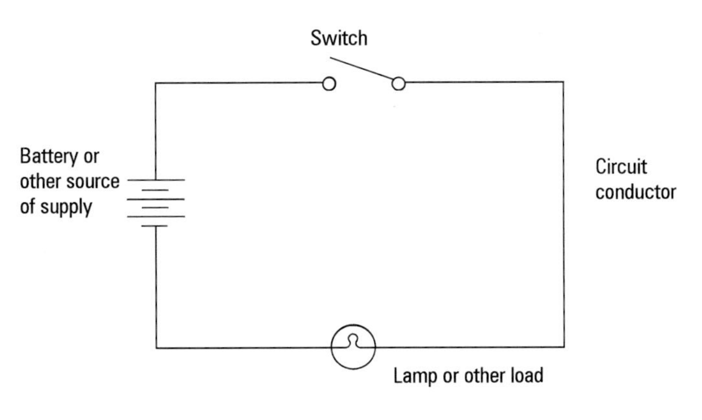


*Figure 3.6*

It is not always practical to have only one load on a circuit. A lighting circuit in the home, for example, may have several lights on a single circuit.

There are two basic types of circuit, series and parallel.

## Series circuits

A circuit connected in series is, as it implies, a number of loads connected one after the other.

If we use a circuit diagram similar to Figure 3.6 but include two lamps, this would look like Figure 3.7.

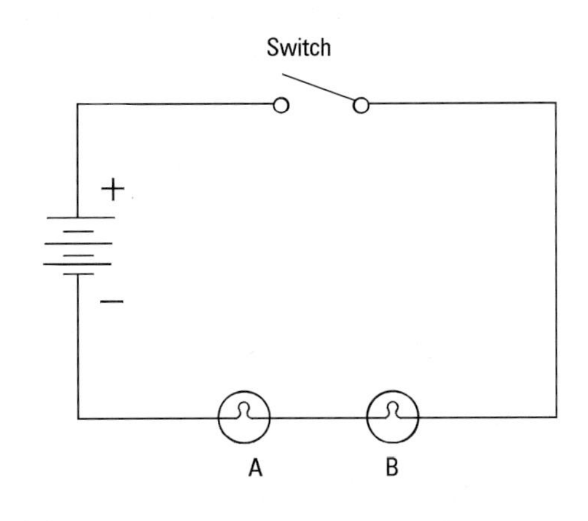


*Figure 3.7*

As both lamps are connected as a single conductor the electrons flowing through **A** must be the same quantity as those flowing through **B**. This means that the current must be the same throughout the complete circuit. When the switch is opened the electron flow is stopped and both lamps go out. This also applies when one of the lamps is removed from its holder, for the lamp filament no longer completes the circuit and the other lamp will also go out. Although this type of circuit has applications it is clearly not used for lighting circuits in the home. If it were then all lights on the circuit would have to be on at the same time, and if one light was switched off all of the others would go out.

## Parallel circuits

When circuits are connected in parallel they can at first appear to be more complex, but wiring techniques have made them straightforward to install and they are far more useful than series circuits. If we adapt Figure 3.6 so that this time there are two lamps in parallel we could end up with the circuit as shown in Figure 3.8.

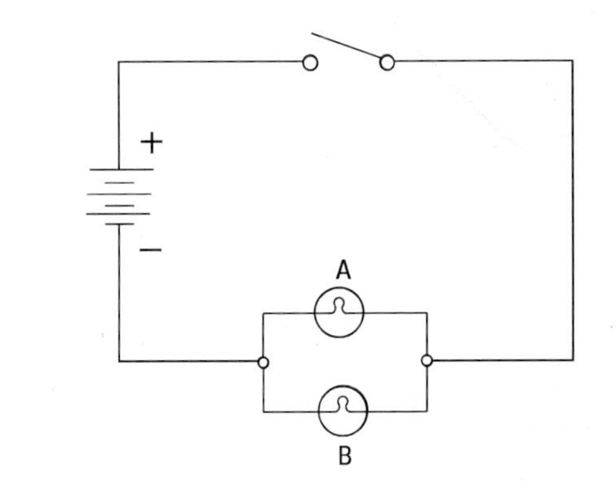


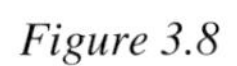

*Figure 3.8*

This could also be drawn, as in Figure 3.9 which is exactly the same.

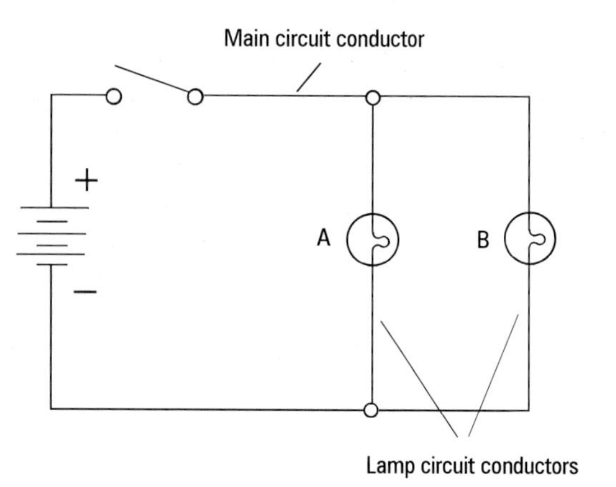


*Figure 3.9*

Now let's see what we have. Each lamp requires its own flow of electrons, independent of the other, but the main circuit conductor has to provide enough for each. If we talk about this in currents, and we assume each lamp is the same, the lamp circuit conductors A and B will both be carrying the same current, but the main circuit conductor will have to carry currents A and B together. If the switch is opened this will stop the current flowing in the main circuit conductor, so both lamps will go out. It is possible though, with parallel circuits, to have each lamp with its own switch, as in Figure 3.10.

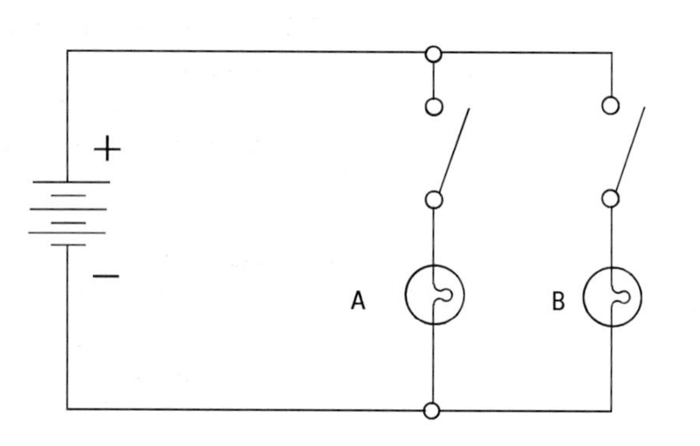


*Figure 3.10*

Now if switch **A** is opened the current ceases to flow in that branch but continues in branch **B** all the time that switch is closed.

Wiring circuits in this way makes them far more practical, and this type of circuit has many more uses than the series circuit.

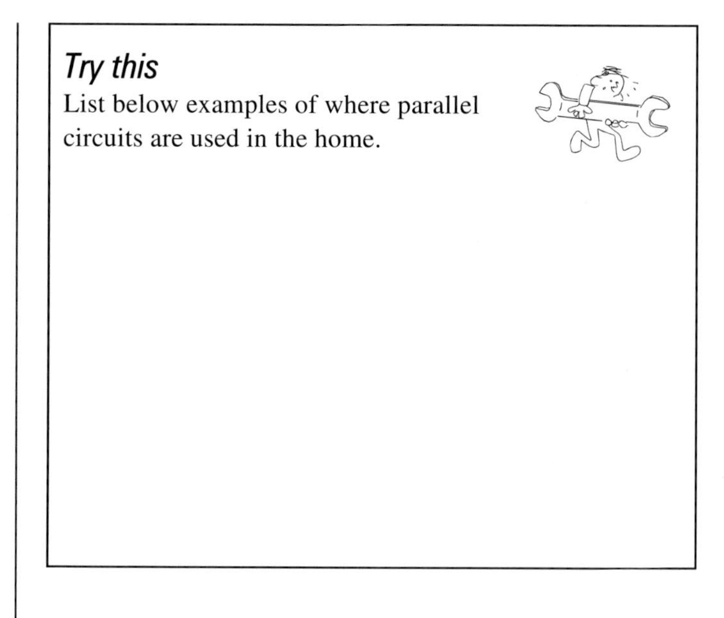

### *Try this*

List below examples of where parallel circuits are used in the home.

### *Points to remember*

## Basic electrical science

All atoms consist of two main component parts. The central nucleus contains the protons and neutrons, and the orbiting satellites are called ____________

Protons are charged positively, neutrons are neutral and electrons are charged ____________

A circuit is a continuous length of ________________ ________________ which starts and ends at the different poles of a source of supply.

The simplest form of control device in a circuit is a ________

The two basic types of circuit are ____________ and ____________

Draw and label below a simple example of each of the two types of circuit.

# Part 2

## Conductors and insulators

Up to now we have been looking at things in broad terms – we will now look at them in more detail.

We have seen that good conductors, such as copper, have free electrons in their atoms. However, other materials will also conduct electricity, but not necessarily as well. Good conductors will pass current with the minimum of effort whereas other materials require greater pressure to be put on them. This pressure comes in the form of voltage, and it is this force that moves the electrons all together in one direction. This is why it is called the **electromotive force (e.m.f.)**. If the pressure is applied at a high enough level, almost any material can be broken down into a conductor.

Materials that will not conduct electricity at normal pressures are called insulators. Often the thickness of an insulator is related to the voltage it is designed to be used for. An example is the insulators for the cables that are carried on pylons across the country (Figure 3.11).

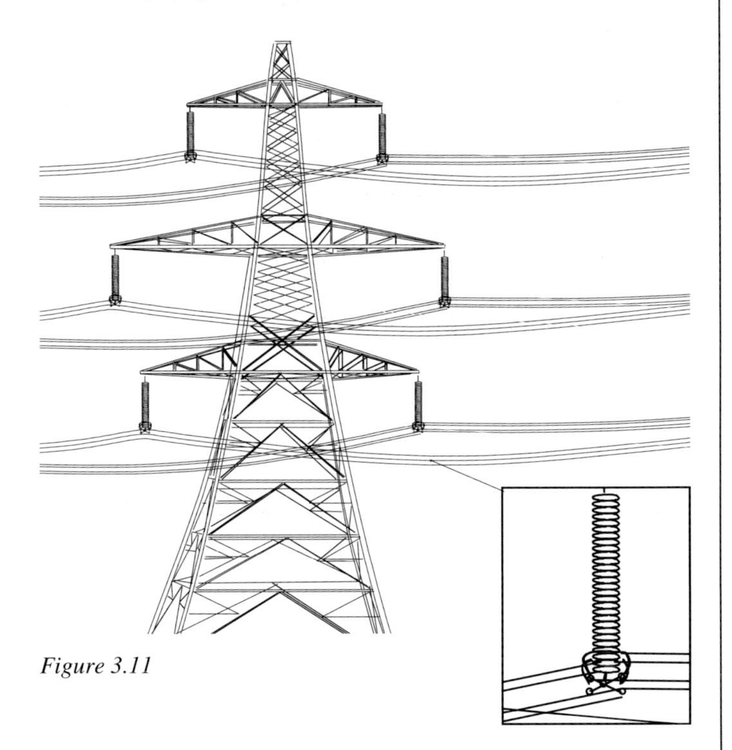

*Figure 3.11*

Air is used to insulate the cables between pylons, and glass or ceramic insulators are used to keep the cables away from each other and the metal structure. The glass insulators are in single units stacked up on each other to provide the required amount of insulation. In general, each glass insulator is used for every 11 000 volts on the line. This means that a 33 000 V line will have a string of at least three insulators. Voltages of up to 400 000 V are transmitted in this way, and you can usually identify these by the long string of insulators at each pylon.

## Resistance

If we are not careful we tend to classify materials as being either conductors or insulators. This would not be true, for many materials will conduct electricity under normal circumstances, but not well enough for use as a cable conductor. These materials, however, often have particular uses of their own. Carbon is a good example of this, for although it can conduct electricity it is not as good as copper or aluminium; Figure 3.12 shows a carbon resistor. There are times when the voltage or the current has to be reduced across parts of circuits in electronics, and materials such as carbon are ideal for this, for they resist the flow of current and can create these changes where required.

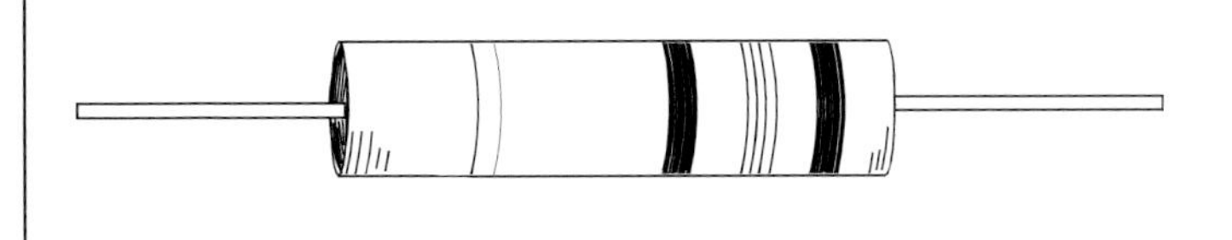

*Figure 3.12 Carbon resistor*

All materials offer some resistance to current flow, even copper and silver. Good conductors, such as these, offer very little resistance under normal circumstances. There are, however, three factors that can change this. They are:

- the length of the conductor
- the temperature of the conductor
- the cross-sectional area of the conductor

The first factor is the length of the conductor. The longer the conductor the further the current has to flow, so the resistance is greater. For example 100 m of a single cable is found to have a resistance of 1 ohm. This means that 200 m of the same cable will have a resistance of 2 ohms.

The second factor is temperature. The higher the temperature the greater the resistance to current flow. Similarly, the lower the temperature the lower the resistance and the greater the conductivity. Carbon is an exception to this, as it has a negative temperature reaction.

The third factor is the cross-sectional area of the material. A thin conductor will have a greater resistance to current flow than a thicker conductor. As most of the conductors we use are round, and the measurements we take from them are the diameters, care must be taken when working out calculations. Remember that the formula for working out the c.s.a. is (Figure 3.13)

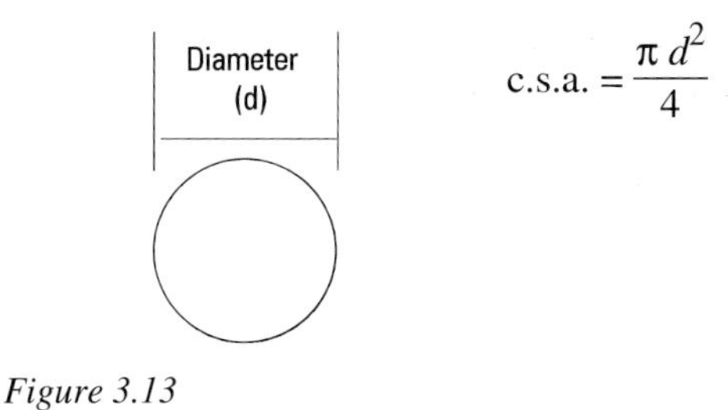


$$\text{c.s.a.} = \frac{\pi d^2}{4}$$

*Figure 3.13*

So if you double the diameter of a conductor you increase its area by a factor of 4. (If you would like to prove this to yourself, work out the c.s.a of two cables 2 mm and 4 mm in diameter and check that one answer is four times the other.) This means that a conductor with a resistance of 1 ohm is twice the diameter of one with 4 ohms, assuming they are of the same material and length.

This can get very interesting if we alter both diameter and length.

> ### *Remember*
> **Resistance opposes the flow of current in a circuit.**
> **Resistance is measured in ohms.**
> **The unit symbol is Ω.**
> **The quantity symbol for resistance is *R*.**
> **Resistance changes with conductor length, cross-sectional area and temperature change.**

### *Example*

Let us take an example of a round conductor with a diameter of 1 mm and a length of 1 metre (Figure 3.14). We require another conductor of the same material with the same resistance, but not the same dimensions.

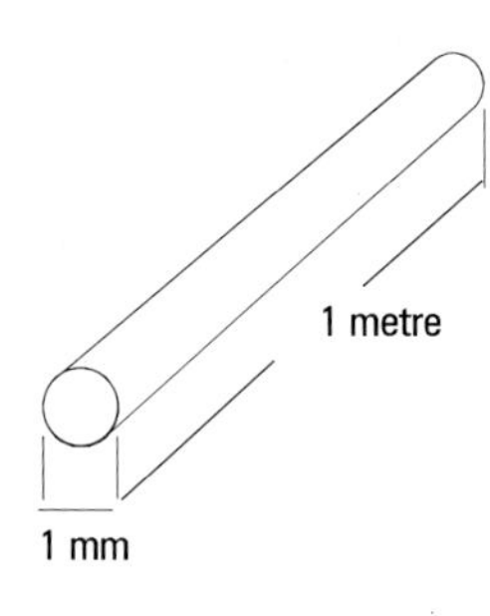


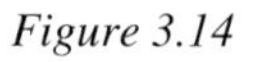
*Figure 3.14*

If we first increase the diameter to 2 mm this means the cross-sectional area has increased by a factor of 4, so if we still had 1 metre in length the overall resistance would be less, in fact only $^1/_4$ of what it was before. To overcome this the length will need to be increased to 4 metres, and then the resistance will be the same as before.

> ### *Try this*
> 
> A conductor has a diameter of 2 mm and a length of 2 metres. A similar conductor is 8 metres long.
> What would the diameter have to be if the resistance stays the same?

To calculate the resistance there are two formulae which take into account the four factors that affect resistance:

- material
- length
- temperature
- cross-sectional area

$$\text{Resistance} = \frac{\text{resistivity} \times \text{length } (l)}{\text{cross–sectional area } (A)}$$

"Material", in this calculation, is the quantity for resistivity: quantity symbol ρ (rho) and unit symbol Ω m, so the formula can be written

$$R = \frac{\rho\, l}{A}$$

The **resistivity** of a material is the resistance of a sample of unit length and unit cross-sectional area. Material that is a good conductor has a low value of resistivity. Material that is a poorer conductor has a higher resistivity.

Some examples of resistivity include

- copper has a resistivity of $17.2 \times 10^{-9}$ Ω m
- aluminium has a resistivity of $28.4 \times 10^{-9}$ Ω m
- nichrome has a resistivity of $1110 \times 10^{-9}$ Ω m

These are the values at a temperature of 20 °C.

So the resistance of a piece of copper 1 metre long and 1 square metre in cross-sectional area is $17.2 \times 10^{-9}$ Ω (or 0.000 000 017 Ω ).

The second formula relates to the resistance after a temperature rise.

$$\frac{R_1}{R_2} = \frac{1 + \alpha\, t_1}{1 + \alpha\, t_2}$$

where

$R_1$ is the resistance at the start temperature
$R_2$ is the resistance at the new temperature
α is the temperature coefficient of resistance
$t_1$ is the start temperature (°C) and
$t_2$ is the new temperature

> ### *Try this*
> 
> What is the resistance of 200 m of 120 mm$^2$ single-core mineral-insulated cable? Resistivity of copper = $17.2 \times 10^{-9}$ Ω m.

The **temperature coefficient of resistance** ($\alpha$) of a material is related to its having a resistance of 1 $\Omega$ at 0 °C to when the temperature is raised by 1 °C.

### *Try this*

A coil has a resistance of 80 ohms at 0 °C. Determine the resistance of the coil at 25 °C. The temperature coefficient of resistance for the coil is 0.004 ohm/ohm °C at 0 °C.

## Voltage

To create the flow of electrons through a circuit pressure has to be applied. As we have seen, this force that moves the electrons is called the Electromotive Force (e.m.f.). This circuit pressure is measured in volts and e.m.f. is usually applied at the source of a circuit. Like pressure in any system, there are parts where the pressure drops. This is referred to in an electrical circuit as a potential difference (p.d.), but it is still measured in volts.

In a series circuit there are potential differences across each component. If we look at the circuit in Figure 3.15 we can see how this applies in practice.

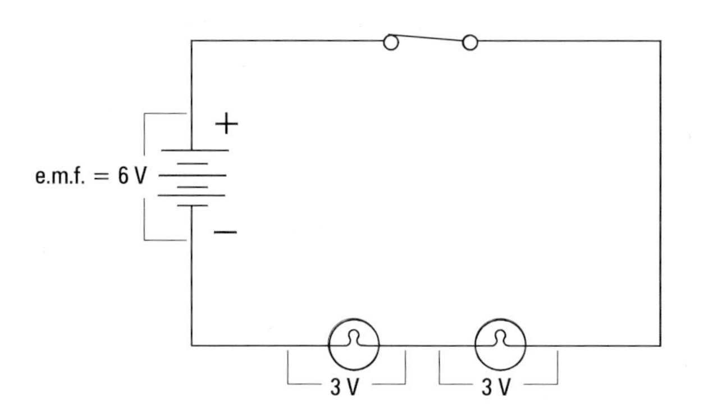


*Figure 3.15* *Series circuit*

If we assume each lamp is the same in voltage and current rating, then when an e.m.f. voltage of 6 V is applied to the circuit this is shared equally across the two lamps. This means that the voltage in the circuit across each lamp is 3 V.

### *Remember*

**Electromotive force (e.m.f.) is measured in volts.**
**The unit symbol is V.**
**The quantity symbol is *V*.**

A circuit containing three identical lamps in series will have the e.m.f. equally divided across the three, as in Figure 3.16

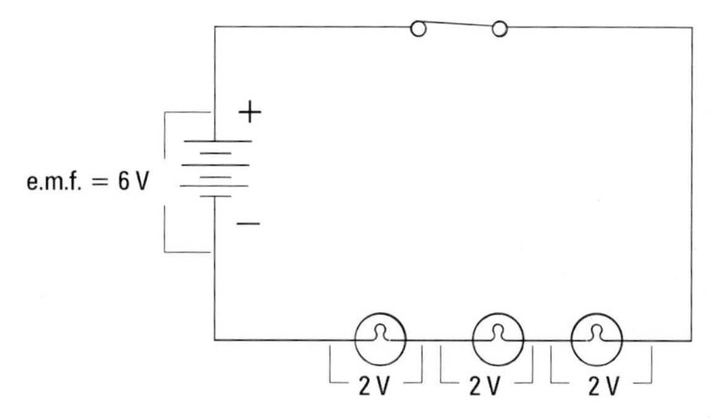

*Figure 3.16* *Series circuit*

In a series circuit the separate p.d.s should add up to the e.m.f.

When considering a parallel circuit the supply e.m.f. should be the same across each branch of the circuit as in Figure 3.17.

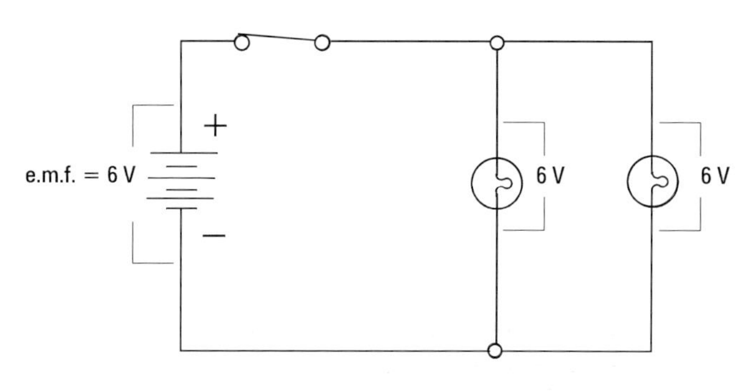

*Figure 3.17* *Parallel circuit*

This is why in household wiring all the lamps are wired in parallel and each lamp is rated and works on the supply voltage.

## Current

We have seen that current is related to the flow of electrons. Now let us see what the relationship is.

The quantity of electricity is measured in coulombs and one coulomb is equal to $6.3 \times 10^{18}$ electrons, that is

1 coulomb = 6 300 000 000 000 000 000 electrons

## Remember

**Quantity of electricity is measured in coulombs.**
**The unit symbol is C.**
**The quantity symbol is $Q$.**

One ampere of electricity is said to flow with the use of one coulomb of electricity in one second.

That is to say that one ampere of electricity flows each time $6.3 \times 10^{18}$ electrons flow past a point in one second.

This can be shown as

$$\text{amperes} = \frac{\text{quantity of electricity in coulombs}}{\text{time in seconds}}$$

$$\text{or} \quad I = \frac{Q}{t}$$

$$\text{or} \quad Q = It$$

## Remember

**Time is measured in seconds.**
**The unit symbol is s.**
**The quantity symbol is $t$.**

When we consider how current is distributed throughout circuits we see that it is an opposite pattern to that of voltage. In a series circuit the current flows through all of the components equally, so it must all be the same.

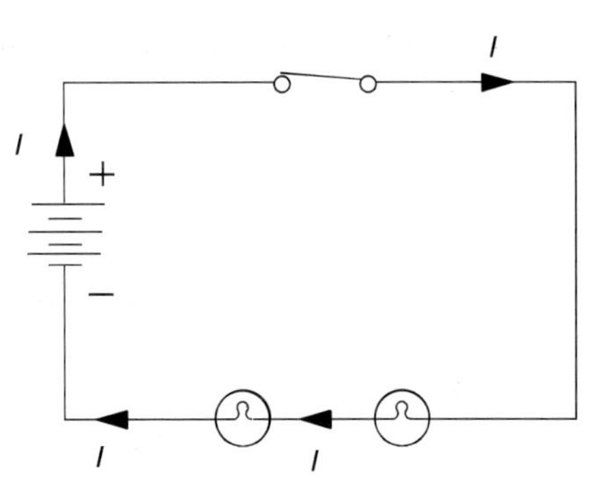


*Figure 3.18* *Series circuit*

All of the currents $I$ in Figure 3.18 will be the same value in a series circuit.

In a parallel circuit the currents in each branch relate to the load connected to that branch. The total current ($I_T$) of a circuit is the sum of the currents in each branch; see Figure 3.19

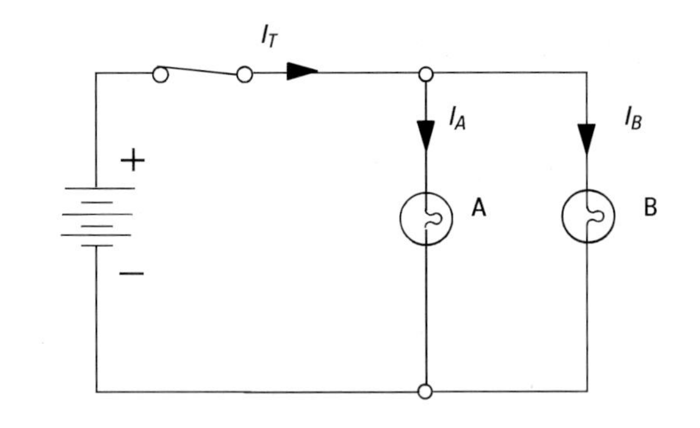


*Figure 3.19* *Parallel circuit*

$I_T = I_A + I_B$ in a parallel circuit.

## Try this

1. Find the resistance of 125 m of 50 mm$^2$ aluminium cable.
   Resistivity of aluminium $28.4 \times 10^{-9}$

2. Determine the resistance of a coil at a temperature of 30 °C if the resistance at 0 °C is 60 ohms. The temperature coefficient of resistance for the coil is 0.004 ohm/ohm °C at 0 °C.

3. How many coulombs of electricity are used when 25 amperes flow for 10 minutes?

4. What current flows when 18 000 coulombs of electricity are used in 20 minutes?

*Points to remember* ◀- - - - - - - - - - - - - - -

## Conductors and insulators

Good conductors will pass current with the minimum of effort. An example of a good conductor material is _________

Materials that will not conduct electricity at normal pressures are called _________________

What is used to insulate cables between pylons?

What materials are used for the insulators on pylons to keep the cables away from each other and the metal structure of the pylons?

All materials offer some resistance to current flow. Good conductors offer very little resistance under normal circumstances. Name the factors that can change the resistance of a material.

Electromotive force (e.m.f.) is measured in ___________
The unit symbol is __ and the quantity symbol is _____

The quantity of electricity is measured in ___________
The unit symbol is __ and the quantity symbol is _____

# Part 3

## Effects of current flow

There are three main effects of current flow. These are:

- the production of heat
- the chemical effect
- the production of a magnetic field

### The heating effect

Usually when we relate the amount of heat being produced by current flow we use the term "power". Electrical power is measured in watts and calculated from

$$\text{watts} = \text{volts} \times \text{amperes}$$
$$\text{or } P = VI$$

#### *Example*

An electrical appliance connected to a 230 V supply and taking 5 A consumes a power of

$$P = VI$$
$$= 230 \times 5$$
$$= 1150 \text{ watts or } 1.15 \text{ kW}$$

The heating effect may be used as a room heater or to heat a filament up in a lamp so that light is given off. It is often a by-product of other effects, such as in an electric motor, where heat is a loss in the electromagnetic process.

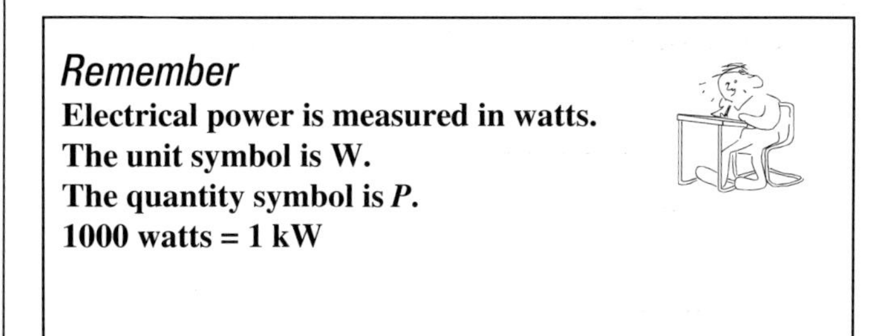

*Remember*
**Electrical power is measured in watts.**
**The unit symbol is W.**
**The quantity symbol is *P*.**
**1000 watts = 1 kW**

### The chemical effect

There are two main ways in which this can be seen.

The first is in batteries and cells (Figure 3.20), where the use of chemicals produces an e.m.f. which creates a current flow in a circuit.

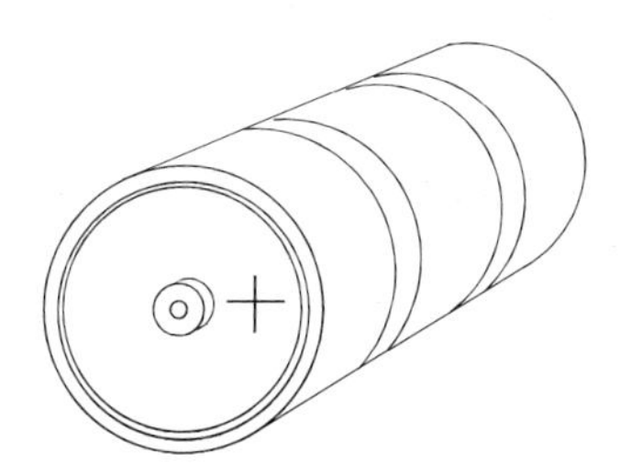


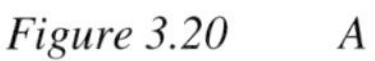

*Figure 3.20  A primary cell*

In secondary cells (Figure 3.21), external currents are used to replenish the chemicals used in the process so that they can be used again (we shall look at this further in Chapter 4).

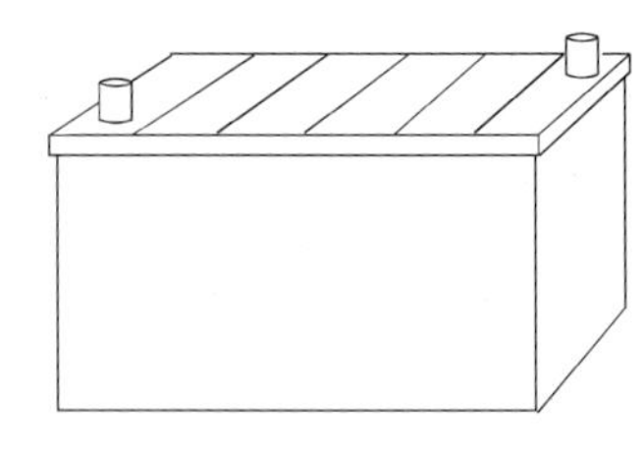

*Figure 3.21 A secondary cell*

The second way in which the chemical effect is used is in the electroplating industry. In this, metallic deposits can be plated on to surfaces by passing a current through electrodes and an electrolyte (Figure 3.22).

The electron flow through the electrolyte carries metal from the negative plate and deposits it on the positive plate. This process is known as electrolysis.

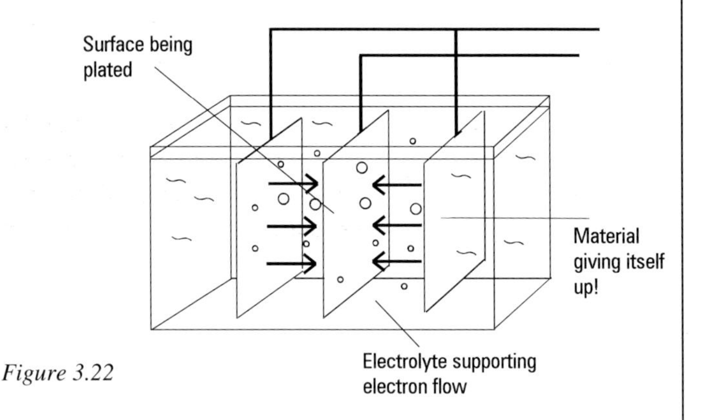


*Figure 3.22*

## The magnetic effect

Of the three effects this is probably the one with the most applications. Whenever a conductor passes an electric current a magnetic field is produced. The direction of the current relates to the direction of the magnetic field (Figure 3.23).

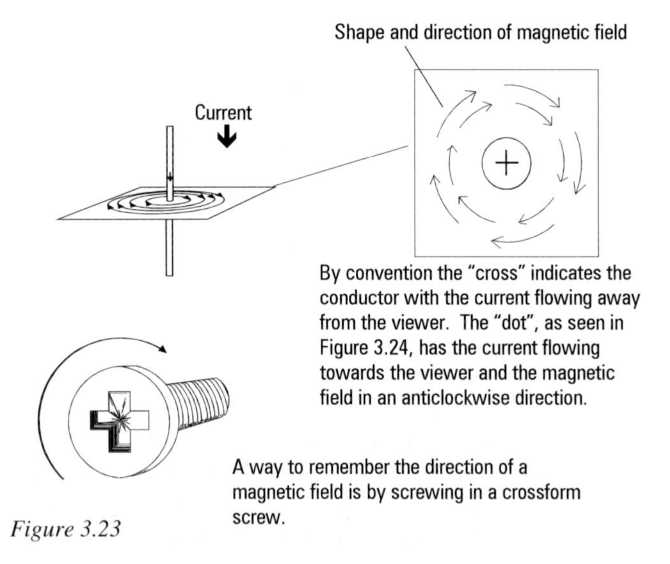


*Figure 3.23*

When two current carrying conductors are placed next to each other they will either repel or attract, depending on the relative direction of the currents (Figure 3.24).

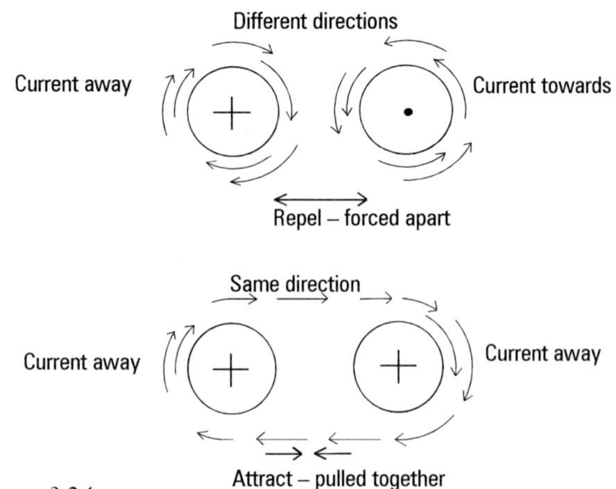

*Figure 3.24*

Electromagnetic fields will also interact with permanent magnetic fields to create movement. The magnetic field strength of the current-carrying conductor is directly proportional to the current flowing in it. Measuring devices use this principle. The magnetic strength of the conductor can be increased by winding it into a coil. The meter that uses this principle is the moving coil meter. A simplified version of this is shown in Figure 3.25.

This type of meter can be used to measure voltage, current or resistance.

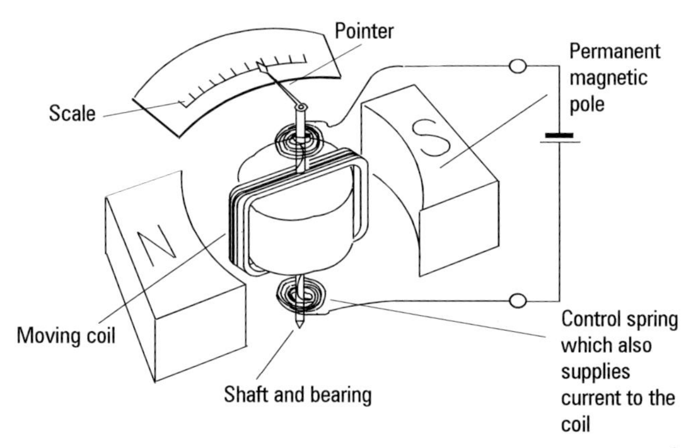

*Figure 3.25 Moving coil meter*

### *Try this*

Look around you and list the devices that use the electromagnetic effect of an electric current.

## Measuring voltage

To measure voltage a meter with the correct type and range should be selected. If a multirange instrument is to be used the correct setting must be selected. On voltages over 50 V a.c. or 120 V d.c. special probes conforming to the Health and Safety Executive Guidance Note GS38 should be used (Figure 3.26). These protect the user from the possibility of electric shock, but care must still be taken.

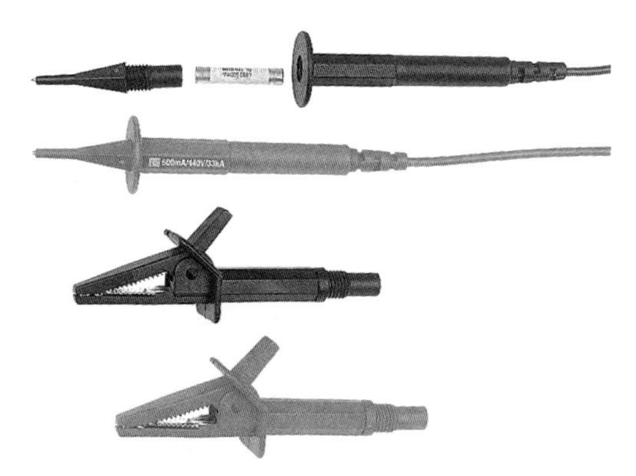

*Figure 3.26* *Fused test prods*

Where possible use meters that can be connected when the circuit is switched off.

Voltage readings may need to be made anywhere across a supply or circuit load. In a series circuit there is going to be far more variation than in a parallel circuit. The correct + and – connections must be used (Figures 3.27 and 3.28). This is known as the correct polarity.

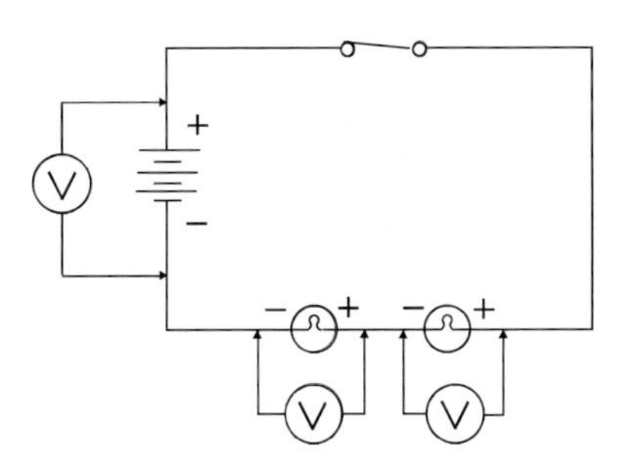

*Figure 3.27*

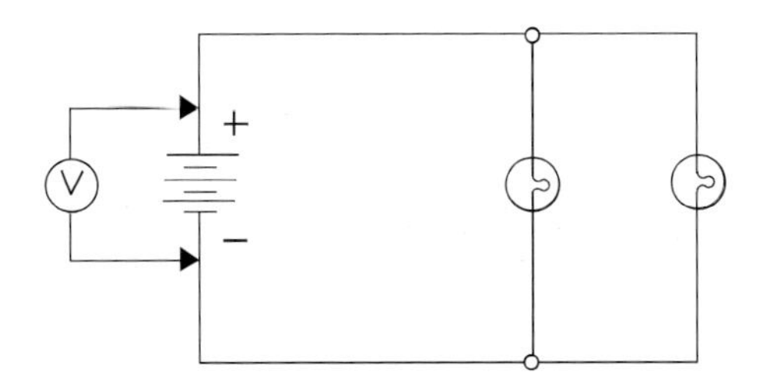

*Figure 3.28* *The voltage across the two lamps should be the same but it can be checked by connecting the voltmeter across them.*

**Safety**

**Use the correct test probes which are designed for your safety.**

## Measuring current

Unlike the voltmeter, an ammeter cannot just be placed across a load to measure the current. An ammeter is a flow meter, so it must be connected into the circuit conductors for the current to flow through it (Figure 3.29). This means that circuit conductors have to be disconnected and the meter fitted in their place. Safety is very important and the circuit should always be OFF before any alterations are made to it. Breaking a circuit that is carrying current can create an arc, which can cause serious burns.

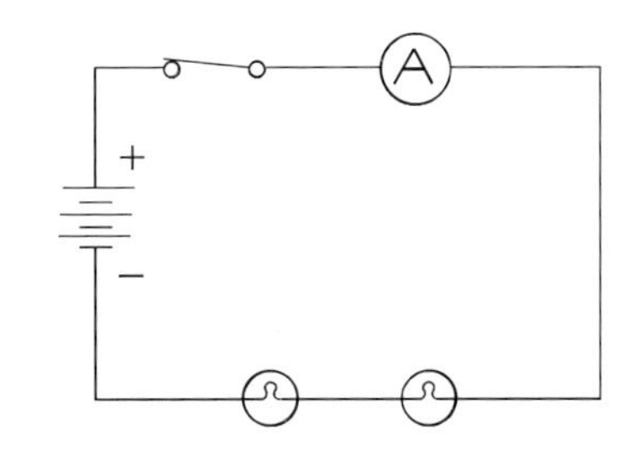

*Figure 3.29* *Ammeters are always connected in series with the load they are measuring*

In a series circuit one ammeter will measure all of the current flowing in the circuit, which will be the same for each load.

In parallel circuits it depends which current reading is required as to where, or how many, ammeters are connected.

The reading on each meter will be related to the load of that part of the circuit. The total circuit current may be measured on another meter connected, as shown in Figure 3.30, or it may be calculated by adding the readings of the ammeters in each branch, as in Figure 3.31.

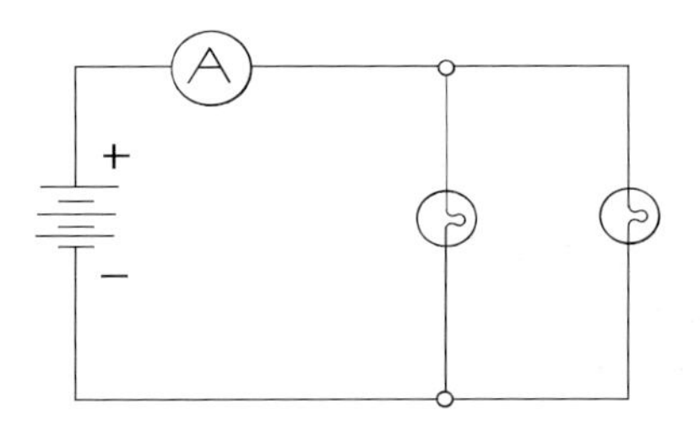

*Figure 3.30*

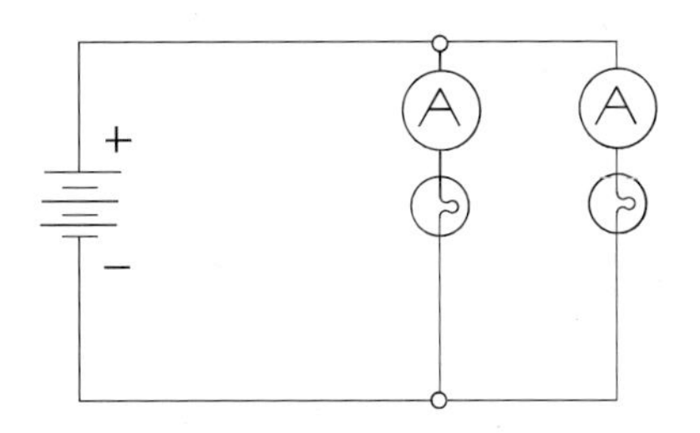

*Figure 3.31*

**Safety**

**The circuit should be off before any alterations are made to it. Breaking a circuit that is carrying current can create an arc which can cause serious burns.**

*Points to remember* ◀- - - - - - - - - - - - - - - -

## Effects of current flow

The three main effects of current flow are:

Electrical power is measured in _________
The unit symbol is __ and the quantity symbol is ____

In the process called electrolysis the _______________ _______________ through the electrolyte carries metal from the negative plate and deposits it on the positive plate.

When a conductor passes an electric current a___________ field is produced.

When two current-carrying conductors are placed next to each other they will either attract or repel depending on the ____ ___________________________________________

What type of meter can be used to measure voltage, current or resistance?

When measuring voltages of over 50 V a.c. or 120 V d.c. special _____________ conforming to Health and Safety Executive Guidance Note GS38 should be used. Whenever possible, use meters that can be used when the circuit is switched off.

Complete the following:

| | Volts | Amperes | Watts |
|---|---|---|---|
| 1. | 230 | 5 | |
| 2. | 110 | 2 | |
| 3. | | 13 | 520 |
| 4. | 80 | | 1000 |
| 5. | 230 | | 3000 |
| 6. | 50 | | 950 |
| 7. | | 10 | 2500 |
| 8. | 50 | 6 | |

# Part 4

## Relationship of voltage, current and resistance

The current that flows in a circuit is directly related to the pressure applied on the circuit by the voltage and the resistance that a circuit offers.

This relationship can be expressed by

$$\text{current} = \frac{\text{voltage}}{\text{resistance}}$$

$$\text{or} \quad I = \frac{V}{R}$$

This is often referred to as Ohm's Law.

### *Example:*

A circuit supplied with 200 V and a load resistance of 10 Ω will have a current flowing of

$$I = \frac{V}{R} = \frac{200}{10} = 20 \text{ A}$$

### *Try this*

1. A load has a resistance of 25 Ω and is supplied with 230 V. What will the circuit current be?

2. A circuit is supplied with a voltage of 110 V and has a load resistance of 22 Ω. What will the circuit current be?

There are occasions when the current is known and either the voltage or resistance needs to be calculated. In these cases the formula has to be re-arranged or transposed.

$$R = \frac{V}{I}$$

and $V = I\,R$

Make sure that you know these formulae and can transpose them at any time.

### Example

When a potential difference of 10 V is applied the current flowing through a resistor is 0.05 A. What is the value of the resistance?

$$R = \frac{V}{I}$$

$$R = \frac{10}{0.05}$$

$$= 200 \text{ ohms}$$

### Example

We know that a heater in a circuit has a resistance of 20 Ω and that it takes 11.5 A. What is the voltage supplied to the circuit?

$$V = I\,R$$

$$V = 11.5 \times 20$$

$$= 230 \text{ volts}$$

It is very important to use this information within circuits, for the values of current, voltage and resistance are often required.

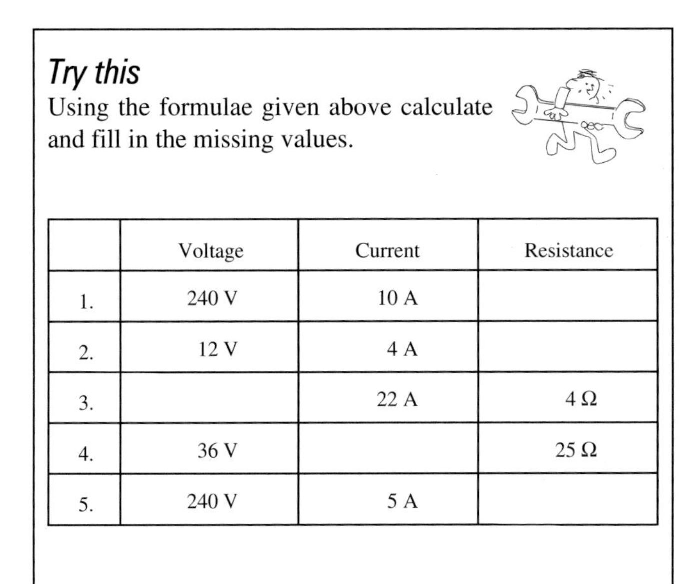

### Try this

Using the formulae given above calculate and fill in the missing values.

| | Voltage | Current | Resistance |
|---|---|---|---|
| 1. | 240 V | 10 A | |
| 2. | 12 V | 4 A | |
| 3. | | 22 A | 4 Ω |
| 4. | 36 V | | 25 Ω |
| 5. | 240 V | 5 A | |

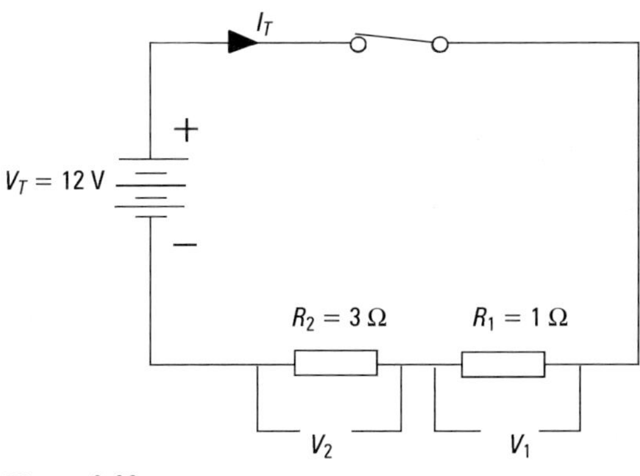


*Figure 3.32*

In the circuit in Figure 3.32 the loads are shown using the general BS circuit symbol for resistors.

Given that the total voltage to the circuit in Figure 3.33 is 12 V and the resistors have values of 1 Ω and 3 Ω, the total circuit current and the potential differences (voltage drops) across the resistors can be calculated.

### Example

In a series circuit the total resistive path to the current flow is found by adding each of the resistors together. In this circuit the total resistance ($R_T$) *is*

$$\begin{aligned} R_T &= R_1 + R_2 \\ &= 1\ \Omega + 3\Omega \\ &= 4\ \Omega \end{aligned}$$

We needed this value before the total circuit current could be calculated.

Total current

$$\begin{aligned} I_T &= \frac{V_T}{R_T} \\ &= \frac{12}{4} \\ &= 3\text{ A} \end{aligned}$$

Notice that to calculate total current the total voltage and total resistance must be used.

Now we can calculate the p.d.s across $R_1$ and $R_2$.
The current through each will be the same as the total $I_T$ which is 3 A.

Using $V = I\,R$ we first calculate for $V_1$.

$$\begin{aligned} V_1 &= I_1 \times R_T \\ &= 3 \times 1 \\ &= 3\text{ V} \end{aligned}$$

Note that because we only need the voltage across $R_1$ the resistance used is $R_1$.

The voltage across $R_2$ can be calculated in two different ways, but the answer should be the same. First we can do it in a similar way to $R_1$, but using the value of $R_2$.

So $V_2 = I_2 \times R_2$
$= 3 \times 3$
$= 9$ V

Or we could have taken the 3 V calculated across $R_1$ and subtracted this from $V_T$.

In this case $V_2 = V_T - V_1$
$= 12 - 3$
$= 9$ V

The answer is the same and it is acceptable to do it either way.

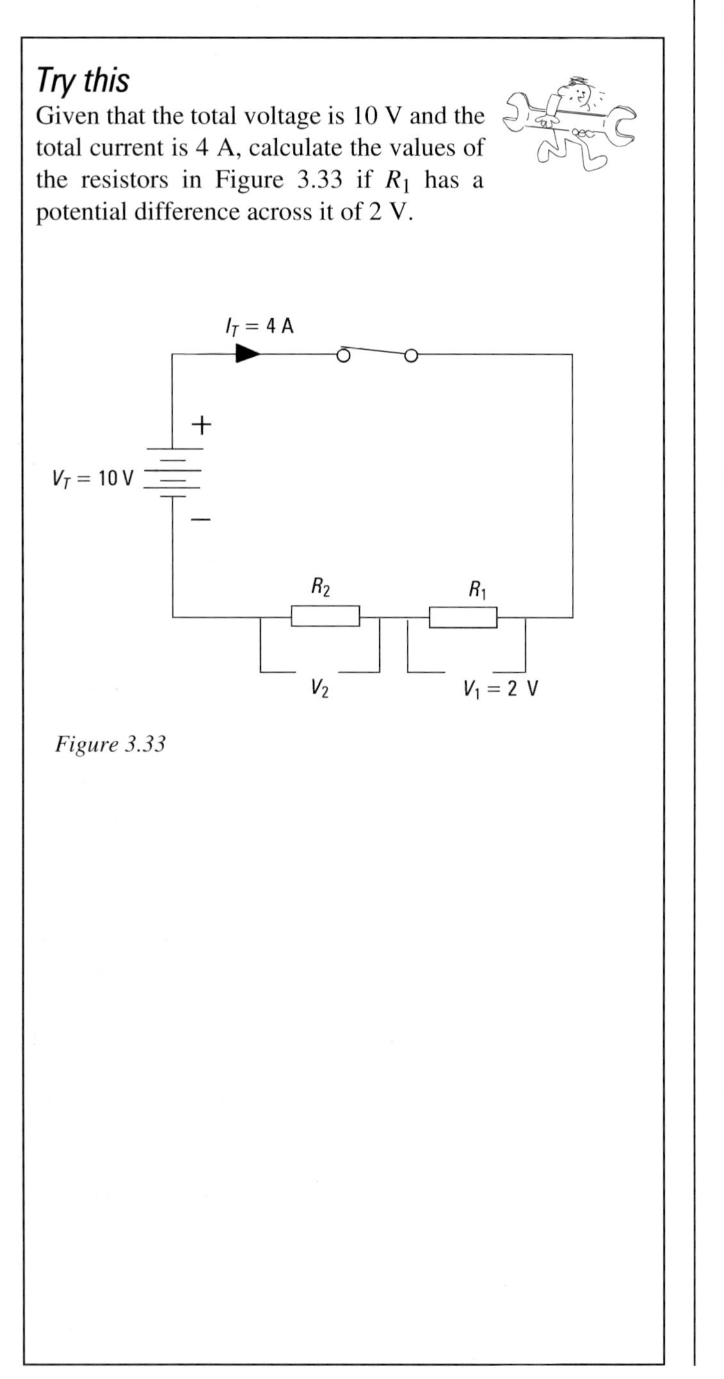


Figure 3.33

## Power

When resistors are used in circuits they may consist of large coils of wire or very small components, such as used in electronics. Very often their physical size is directly related to the power the resistor has to dissipate. Earlier we saw that power was equal to the voltage times the current, $P = V \times I$. If we consider what we have seen with the relationship between voltage, current and resistance, there are two other ways of calculating power.

$P = I^2 R$ and

$P = \frac{V^2}{R}$

Both of these methods use the value of the resistor to help to determine the power.

If we now consider two circuits (Figures 3.34 and 3.35), each with 20 Ω resistors in them, we can see that the power rating of the resistor is important.

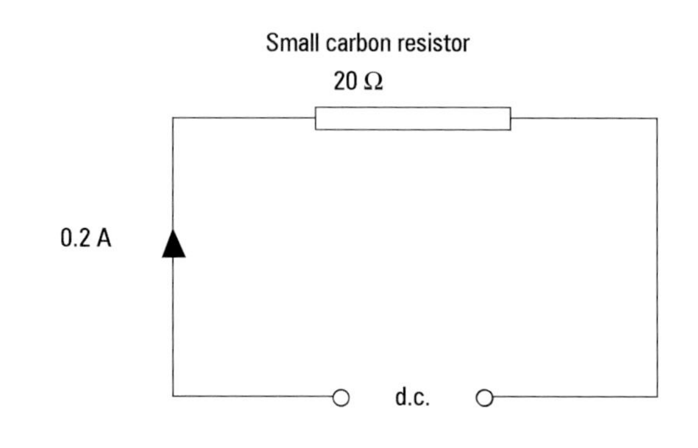


Figure 3.34

$P = I^2 R$
$= 0.2 \times 0.2 \times 20$
$= 0.8$ W

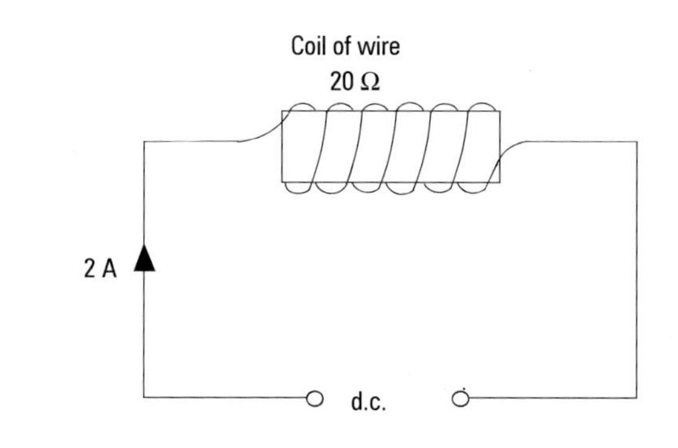


Figure 3.35

$P = I^2 R$
$= 2 \times 2 \times 20$
$= 80$ W

The amount of heat dissipated in each resistor is completely different. If the small carbon resistor was used to replace the coil of wire it would blow up, and if the large coil was used in the electronic circuit it would be physically out of proportion. Therefore the power rating of resistors is important.

Power is also important when considering the rating of cables. A 3 kW immersion heater takes far more current than a 100 W lamp when they are connected to the same 230 V supply.

If we transpose the formula $P = IV$ so that $I$ is determined, we end up with

$$I = \frac{P}{V}$$

The current for a 3 kW heater is:

$$I = \frac{P}{V}$$
$$= \frac{3 \times 1000}{230}$$
$$= 13.04 \text{ A}$$

whereas the 100 W lamp has a current of

$$I = \frac{P}{V}$$
$$= \frac{100}{230}$$
$$= 0.43 \text{ A}$$

This difference in current affects the selection of the cable and protection device of the circuit.

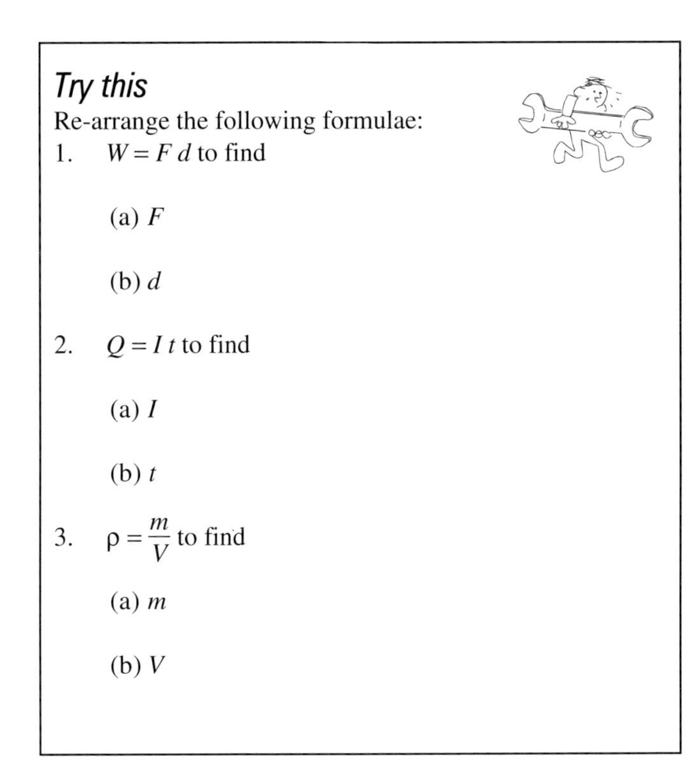

*Try this*

Re-arrange the following formulae:

1. $W = F\,d$ to find

   (a) $F$

   (b) $d$

2. $Q = I\,t$ to find

   (a) $I$

   (b) $t$

3. $\rho = \frac{m}{V}$ to find

   (a) $m$

   (b) $V$

## Wattmeter

The power in a circuit can be measured using a voltmeter and an ammeter (Figure 3.36).

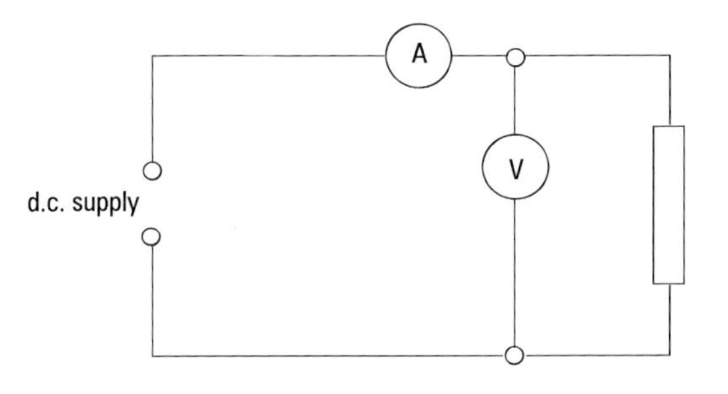


*Figure 3.36*

The readings of each are multiplied together to give the power reading. This can be carried out in one instrument which measures both voltage and current and displays the result in watts (a wattmeter) (Figure 3.37).

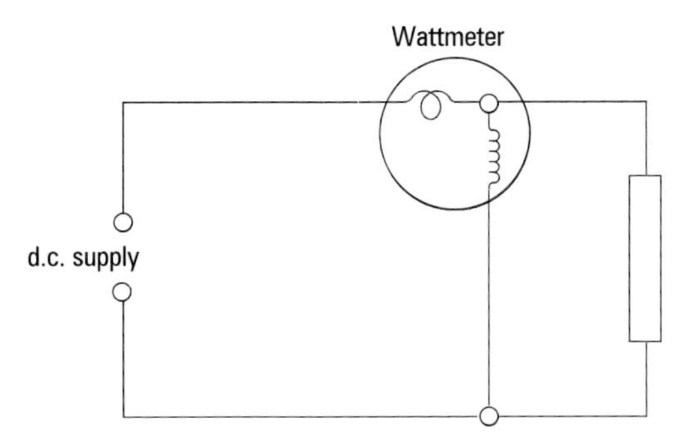


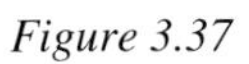

*Figure 3.37*

The connection of these meters has to be carried out with care, for if they are connected incorrectly the meter may burn out. The current coil must be connected in series, the same as a separate ammeter, and the voltage coil in parallel, the same as a voltmeter.

On some wattmeters links have to be connected between the current and voltage coils; on others this is already carried out.

*Remember*

| Quantity | Quantity symbol | Measured in | Unit symbol |
|---|---|---|---|
| Current | $I$ | amperes | A |
| Voltage | $V$* | volts | V |
| Resistance | $R$ | ohms | Ω |
| Power | $P$ | watts | W |

*Uo as the nominal supply voltage may be used for V.

### *Try this*

1. In a circuit there are three resistors of 1 Ω, 3 Ω and 2 Ω in series. Calculate the total circuit current and the voltage drop across each resistor if the total circuit voltage is 12 V.

2. What is the power taken by a heater in a circuit connected to a 230 V supply and a resistance of 20 Ω? Give your answer in kW.

3. What is the current taken by a 150 W lamp connected to a 230 V supply?

### *Points to remember*

## Relationship of voltage, current and resistance

The formula referred to as Ohm's Law is

In a series circuit the total resistive path to the current flow is found by ______________ each of the resistors together.

To calculate total current in a series circuit the values of the total ___________ and the total ___________ must be used. In a series circuit the current will be the same through all the resistors.

## Power

Power is equal to the voltage times the ____________

Two other methods of obtaining the power in a series circuit are by using the value of the resistor in the formulae:

$P$ =

$P$ =

The power in a circuit can be measured using a voltmeter and an ammeter or by using a _______________

### *Self-assessment multi-choice questions*

**Circle the correct answers in the grid below.**

1. Electromotive force is measured in
   (a) amperes
   (b) volts
   (c) ohms
   (d) watts
2. A conductor has a diameter of 4 mm and a length of 2 metres. A similar conductor has a diameter of 2 mm. What would the length have to be if the resistance stayed the same?
   (a) 0.5 m
   (b) 1m
   (c) 8 m
   (d) 4 m
3. How many coulombs of electricity are used when 20 amperes flow for 1 hour?
   (a) 12 000 coulombs
   (b) 18 000 coulombs
   (c) 36 000 coulombs
   (d) 72 000 coulombs
4. An electrical appliance connected to a 230 V supply and taking 3 A consumes a power of
   (a) 0.46 kW
   (b) 0.69 kW
   (c) 1.69 kW
   (d) 4.6 kW
5. A circuit supplied with 230 V has a current flowing of 20 A. What resistance does the circuit offer?
   (a) 11.5 Ω
   (b) 115 Ω
   (c) 4.6 Ω
   (d) 460 Ω

### *Answer grid*

1 a b c d
2 a b c d
3 a b c d
4 a b c d
5 a b c d

# 4

# Electrical Sources and Effects

Write down the answers to the following questions.

Examples of materials which make good conductors are:

Examples of materials which make good insulators are:

In what way do the following factors affect the resistance to current flow in a circuit:

- The length of the conductor
- The temperature of the conductor
- The cross-sectional area of the conductor

What are the three main effects of current flow?

Which formula expresses the relationship of voltage, current and resistance?

**On completion of this chapter you should be able to:**

- state the properties of a magnetic field
- describe how a single-phase a.c. is generated
- calculate the voltage and current outputs of a transformer
- explain the relationship between the phases of a three-phase supply
- determine the relationship between voltages in star and delta connected supplies
- describe how a cell produces electricity
- identify primary and secondary types of cell
- describe correct battery charging procedures
- complete the revision exercise at the beginning of the next chapter

# Part 1

## Production of an electromotive force (e.m.f.)

There are a number of ways to create an e.m.f. They all involve converting one of the other forms of energy into electrical energy.

Alternators convert magnetic energy into electrical energy.

Cells convert chemical energy into electrical energy.
A chemical change takes place in a cell that produces electricity.

The thermocouple converts heat energy into electrical energy, and this process is known as the Seebeck effect. A thermocouple is used extensively for measuring high temperatures. The thermocouple is a device consisting of wires of two different metals which are connected together to form junctions. When one of the junctions is heated a potential difference will be created between it and the cold junction. This potential difference will be very small, but nevertheless it will be proportional to the temperature difference between the junctions.

Solar cells convert light energy into electrical energy. They are semiconductor devices and are based on the direct conversion from light energy into electricity. They are frequently used for satellite communications or in remote locations.

The piezoelectric effect converts mechanical energy into electrical energy. An example of the use of this effect is in microphones or the hand-held gas ignitor for your gas hob.

### Production of an e.m.f. by magnetic means

There are two main components required to generate a voltage in this way:

- a conductor
- a magnetic field

As soon as one of these is moved relative to the other the electrons in the conductor become excited and start to move. A simple way of showing this is with a coil of copper wire, a bar magnet and a sensitive voltmeter or galvanometer.

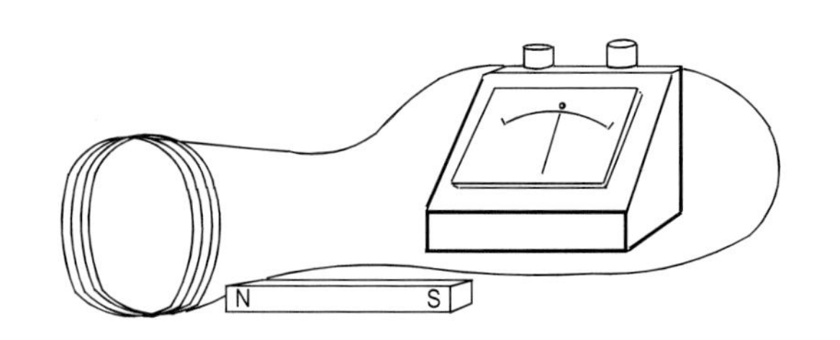


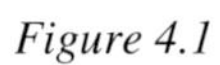

*Figure 4.1*

When the coil is connected to the voltmeter, as shown in Figure 4.1, and the magnet is laying away from the coil, the needle in the meter is in the centre at zero. As soon as the magnet is moved closer to the coil the needle moves. The movement is created by an electromotive force (e.m.f.) – "the force that moves electrons".

To understand why this happens we must look at the magnetic field around a permanent bar magnet (Figure 4.2).

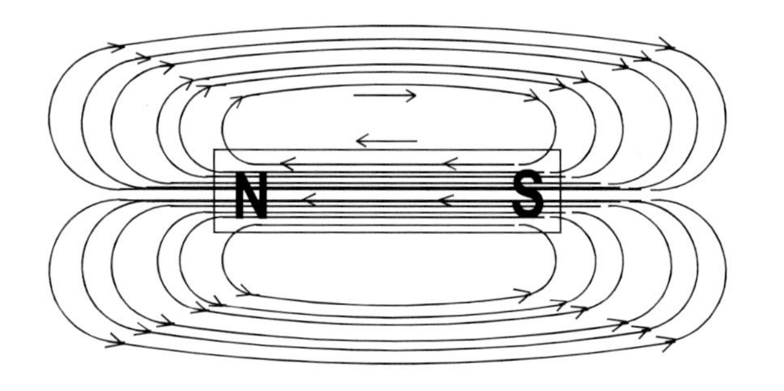


*Figure 4.2* *The magnetic field around a bar magnet*

All of the lines of force come from the North pole, go through the air and back into the magnet at the South pole. They then go through the magnet back to the point from where they started. This circuit of magnetic field is very important, for it demonstrates many of the laws that apply to all magnetic circuits.

These are that magnetic lines of flux

- go outside the magnet from the North pole to the South pole
- return inside the magnet from the South pole to the North pole
- are all complete circuits of magnetism
- never cross one another
- cannot be broken, only reshaped

We need now to go back to our original equipment, as shown in Figure 4.1. When the magnet is lying inside the coil of copper wire, the magnetic field goes right through the wire. When the magnet and coil are stationary the magnetic field is stationary inside the wire. However, as soon as the magnet or coil is moved the magnetic field moves through the conductor, forcing the electrons to flow. If the magnet is moved in the opposite direction, the magnetic field moves the electrons through the conductor in the opposite direction. This in turn makes the needle on the meter move to the opposite side.

So let's just check this. Assume that the North pole is moved into the centre of the coil and that as this is done the needle on the meter moves to the left (Figure 4.3).

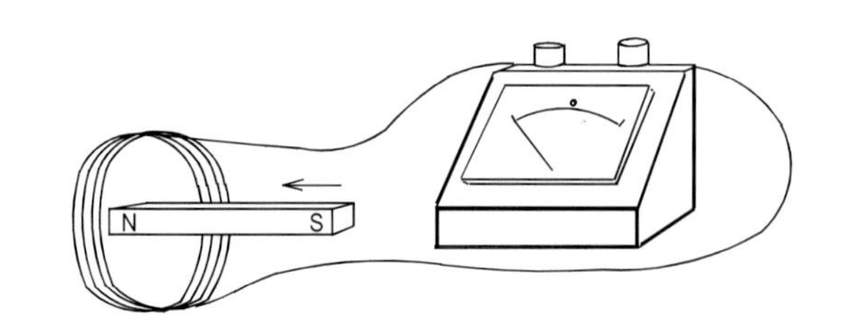


*Figure 4.3* *The magnet going into the coil*

The magnet is then pulled back out of the coil and the needle moves to the right (Figure 4.4).

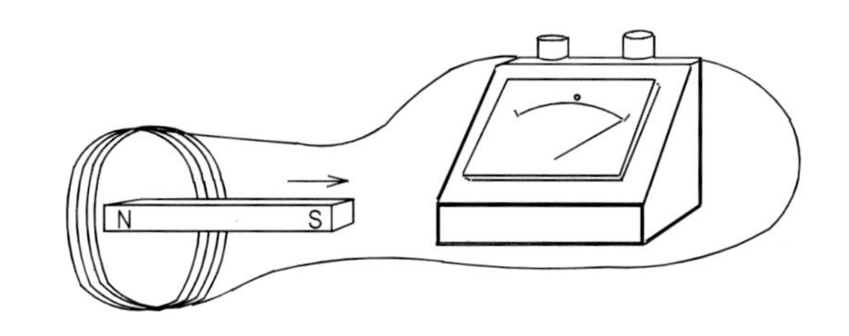


*Figure 4.4* *The magnet pulling back from the coil*

This effect of making the needle reverse can be achieved by passing the magnet right through the coil and out the other side. When the North pole goes through the needle moves to the left as before, but when the South pole goes through the needle swings to the right.

If at any time the magnet becomes stationary the needle goes back to zero.

The voltmeter is measuring a voltage that is produced by the movement of electrons through the coil, indicating that a current, albeit small, is flowing within the coil.

### *Try this*

Name three devices that use the principle of moving magnetic fields within a coil to make them work.

1.

2.

3.

## An alternating current (a.c.)

We have seen what happens when a magnet is moved inside a coil. Now we want to use this theory to make a practical generator to produce alternating current.

The first thing is to have the magnet on the outside and the coil inside. The coil must be fitted to a shaft and bearings so that it can rotate. This means that the connections to the coil must be made through rings (slip rings) that are fitted to the shaft and slide against fixed contacts connected to the meter. As the coil rotates, each of the two long sides goes past the poles in turn. At the instant the coil is as shown in Figure 4.5, both points A and B are cutting through the lines of flux at the maximum rate. This produces a maximum deflection on the meter to the left. As the coil rotates, the e.m.f. drops off until the coil is vertical between the poles. At this point the e.m.f. is zero.

The coil now continues to rotate and the e.m.f. builds up again, but in the opposite direction.

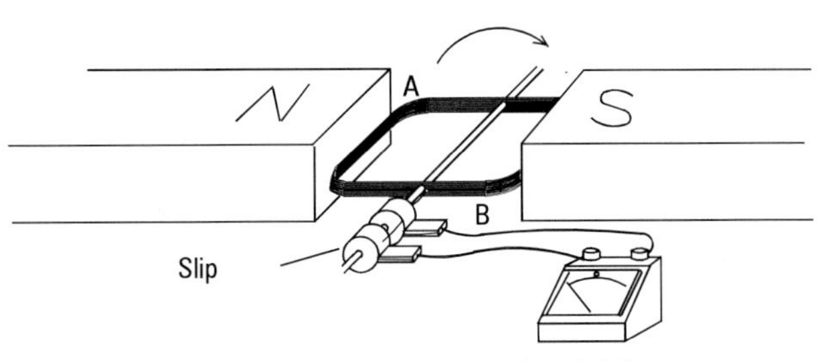


*Figure 4.5*

The effect of rotating the coil continually changes the direction of the current flowing in the circuit.

This is why this system of supply is known as alternating current (a.c.). As there is only one output from this type of generator it is known as a single-phase supply.

The speed at which the coil rotates is very important as this governs the frequency, or cycles per second, of the supply. In the UK and Europe the frequency is 50 hertz (50 Hz). This means that there are 50 cycles each second. The coil in the generator we have looked at has to turn 50 complete revolutions each second to achieve this.

The sequence that the current goes through in one cycle is called a waveform. This is one 360° turn of the coil. The type of wave this produces is a sinusoidal or sine wave (Figure 4.6). This waveform is the same as that of the voltage which appears during the generation of current within the coil.

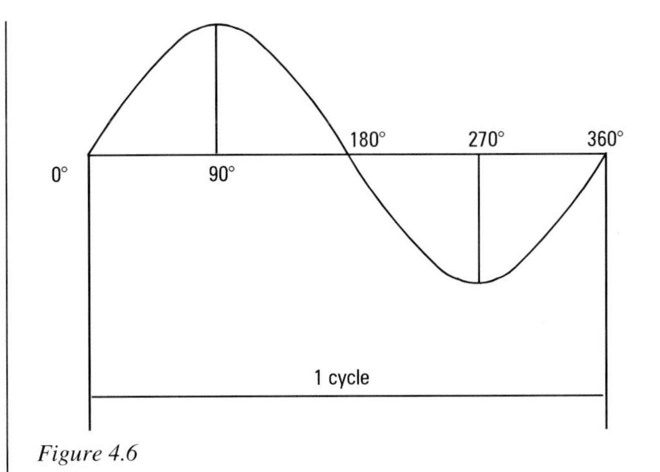


*Figure 4.6*

### *Try this*

1. What is a.c. the abbreviation for?

2. What are slip rings used for?

3. How many degrees does a coil need to rotate to create one complete sine wave?

## Three-phase generators

Although single-phase is used for domestic supplies, it is seldom generated in large quantities. Power stations use generators (or, as we should call them, alternators) that produce three phases.

The principle of a three-phase alternator is similar to that of a single-phase device, but the construction is quite different. This time the generating coils remain stationary while the magnetic field revolves. Although in practice the moving magnet is an electromagnet fed from a d.c. supply, we shall show it as a permanent magnet to keep it simple.

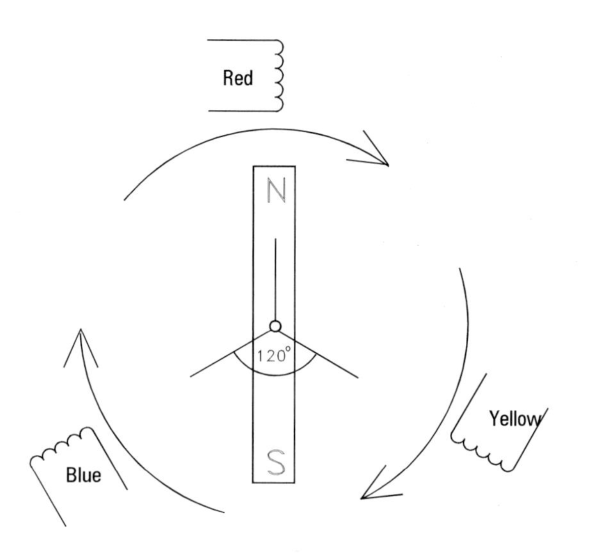


*Figure 4.7*

As the magnet shown in Figure 4.7 rotates past each coil in turn, an e.m.f. is generated in that coil. As the coils are 120° apart, the sine wave generated in each is 120° "out of phase" with the next one. This creates three waveforms starting 120° after each other, as shown in Figure 4.8.

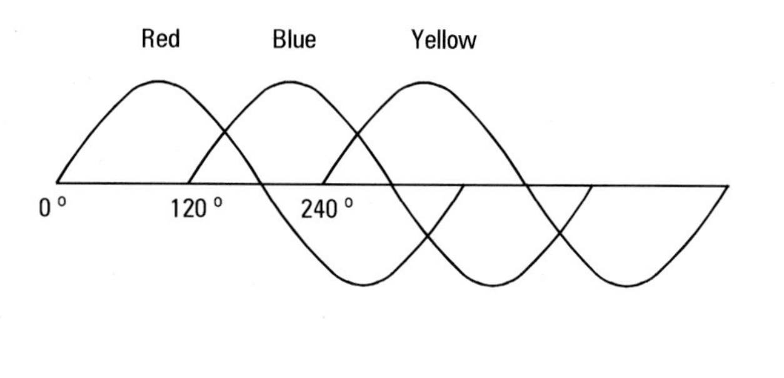


*Figure 4.8*

A three-phase supply can be used in many different ways and can be very adaptable. It is particularly suitable for large power applications.

## The electromagnet

If we take a coil of wire and connect it to a d.c. supply (Figure 4.9) we can see (Figure 4.10) that the magnetic field distribution is very similar to that around a permanent bar magnet. A coil, consisting of a number of turns of wire wound in the same direction, which is capable of carrying a current is called a **solenoid**.

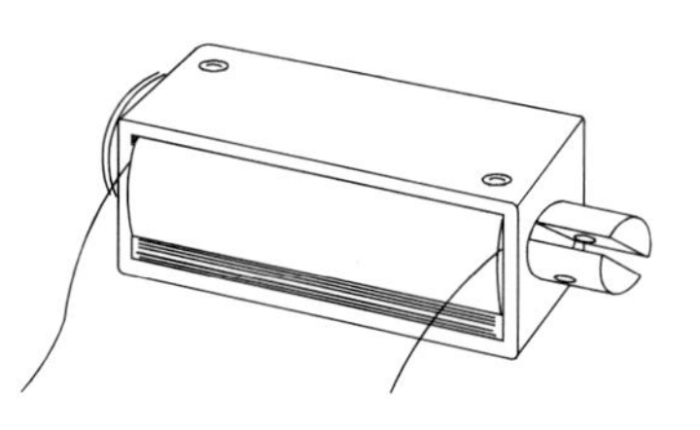

*Figure 4.9* *A solenoid*

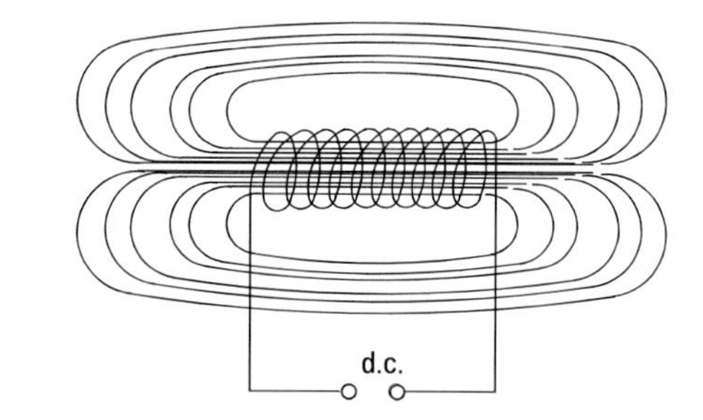


*Figure 4.10* *The magnetic field around an electromagnet*

All of the same rules for the lines of magnetic flux apply as before.

If an iron bar is placed inside the coil, as in Figure 4.11, you can see that the field through the coil is now concentrated within the iron.

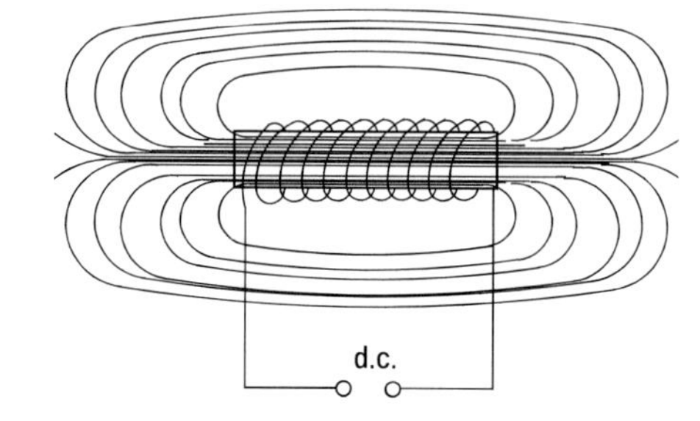


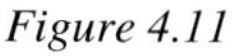

*Figure 4.11*

The iron bar is now similar to a bar magnet in that it has North and South poles (Figure 4.12). Which end of the bar is which depends on the direction of the d.c. supply.

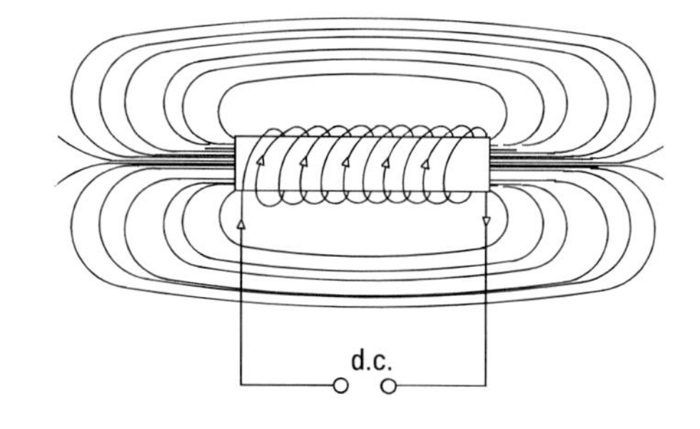


*Figure 4.12*

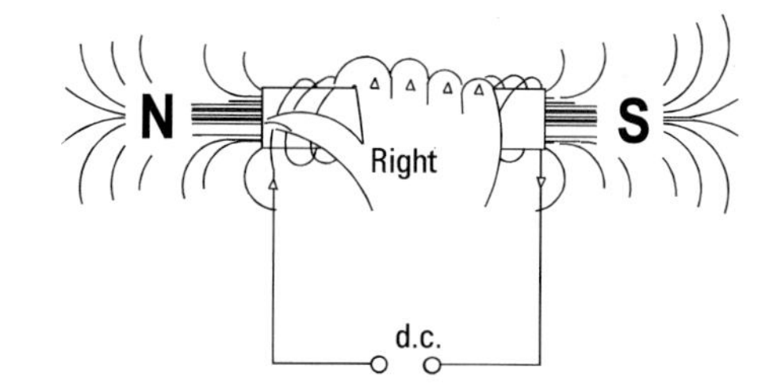


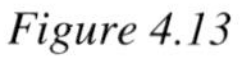

*Figure 4.13*

If your right hand is placed over the coil so that your fingers are pointing in the direction of the current, your thumb points to the North pole (Figure 4.13).

When the d.c. supply is reversed the North and South poles take up positions at the opposite ends of the bar.

> *Remember*
> **Each line of magnetic flux:**
> - goes from North to South poles outside a magnet
> - returns through a magnet from South to North
> - is a complete circuit
> - is independent and does not touch or cross another

> *Try this*
> Mark on Figure 4.14 which is the North pole and which is the South pole of the steel bar in the coil.

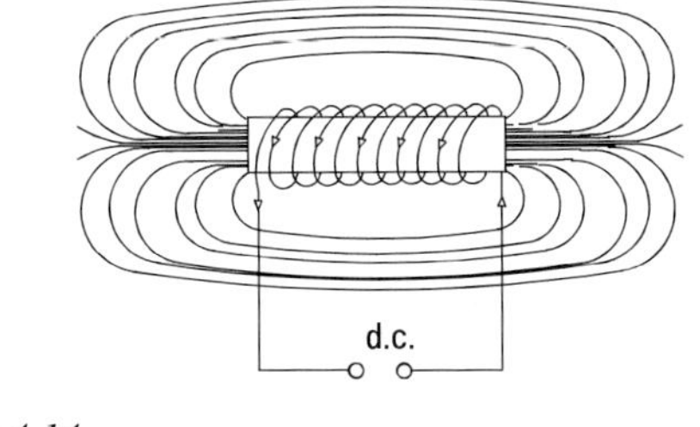


*Figure 4.14*

## *Points to remember*

## Production of an e.m.f.

____________ convert chemical energy into electrical energy.

A ____________ converts heat energy into electrical energy.

____________ convert magnetic energy into electrical energy.

To generate a voltage by magnetic means requires a _______ and a ____________

Magnetic lines of flux go ____________ a magnet from North to South and return ____________ a magnet from South to North. Each line of magnetic flux is a complete ____________ and does not touch or cross another.

What is frequency?

Frequency is measured in ____________ and in the UK the public supply frequency is ____________

A solenoid is

# Part 2

## The double-wound transformer

The double-wound transformer consists essentially of two coils insulated from each other and mounted on a common magnetic core: see Figure 4.15.

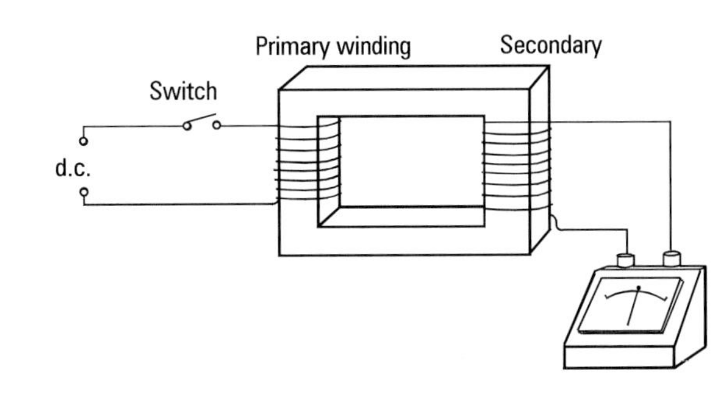


*Figure 4.15*

When the switch is closed a magnetic field is set up in the core from the primary winding. This magnetic field moves through the second coil (secondary winding) as it sets itself up. This movement of magnetic flux is enough to induce an e.m.f. into the second coil and the meter moves to one side. Once the current in the primary reaches a steady state the magnetic flux will be constant. As there is no change in flux it is the same effect as the bar magnet being stationary, so the meter on the second coil goes back to zero. When the switch is opened the magnetic field collapses and is again moving, so an e.m.f. is induced into the second coil. This induced e.m.f. is in the opposite direction as the magnetic field is moving back to zero. The deflection on the meter is in the opposite direction and again returns to zero as the magnetic field stops.

Each time the switch is closed or opened there is a small deflection on the meter. Obviously, if a constant output is required from the secondary coil the supply to the input coil must be a.c. As this automatically switches on and off 100 times each second the output on the second coil is also a.c., as shown in Figure 4.16.

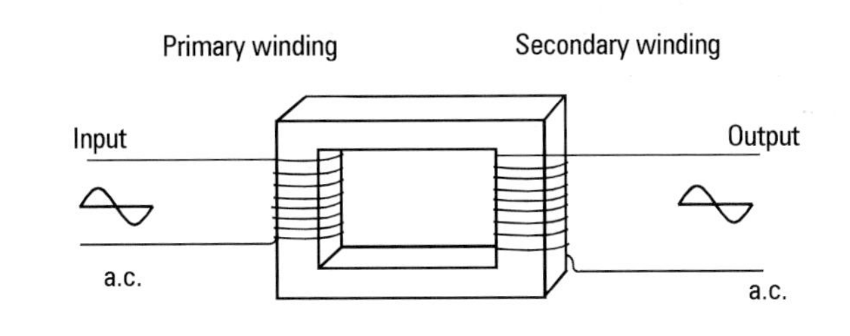


*Figure 4.16*

## Turns ratio

Transformers on supply systems can be "step-up" or "step-down". This means that the voltage coming out of a transformer may be made greater or less than the input.

Let's consider the transformer in Figure 4.17, which has been manufactured with 10 turns on the input winding and 100 turns on the output.

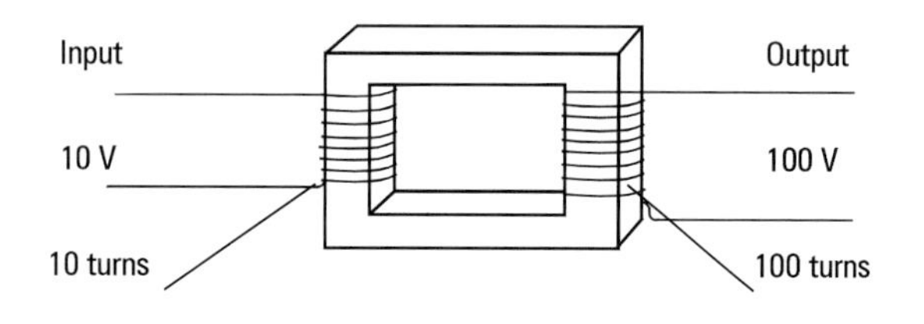


*Figure 4.17*

When 10 volts a.c. is connected to the input winding, 100 volts is available at the output (ignoring any transformer efficiency). This is known as a step-up transformer. If the connections are reversed this now becomes a step-down transformer. If 10 volts is connected to the 100 turn coil, as in Figure 4.18, only 1 volt will be available at the terminals of the 10 turn coil. This means this transformer has a turns ratio of 10:1.

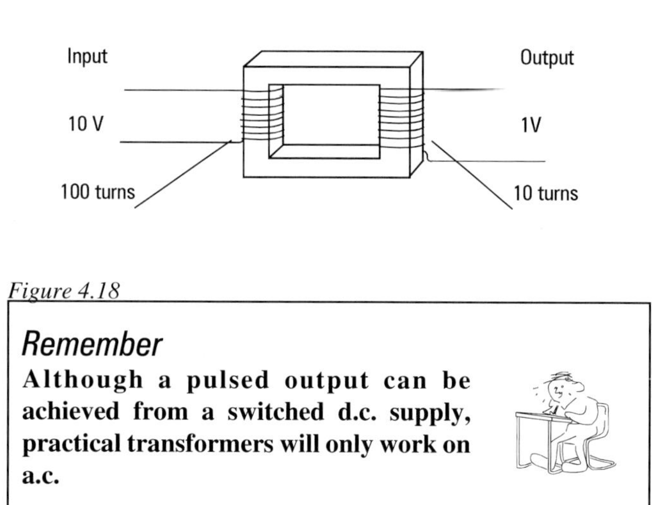


*Figure 4.18*

### *Remember*

**Although a pulsed output can be achieved from a switched d.c. supply, practical transformers will only work on a.c.**

### *Example*

Let's assume that we have a single phase 240 V supply which we need to step down to 12 V for a front door bell, as shown in Figure 4.19. We have a transformer which has 1200 turns on the primary side. How many turns would we need to wind for the output coil?

The ratio is 240:12
or 240/12
or 20:1

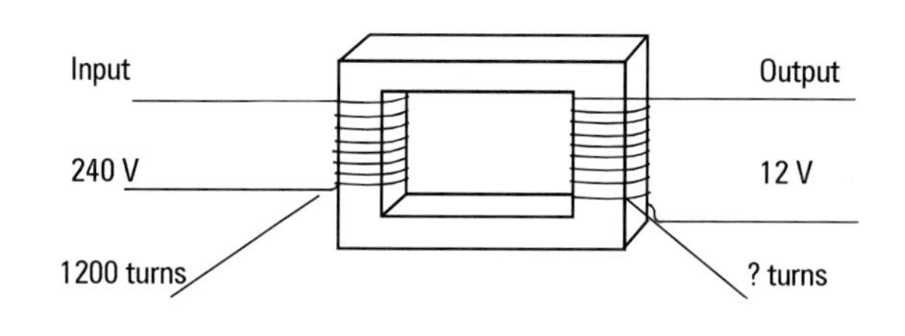


*Figure 4.19*

This means that 1200/20 will give the number of turns on the output side.

$$= \frac{1200}{20}$$

= 60 turns

### *Try this*

A transformer has 960 turns on one winding and 120 turns on the other. This is to be used as a step-down transformer and the input voltage is to be 240 volts, as shown in Figure 4.20. What will the output voltage be?

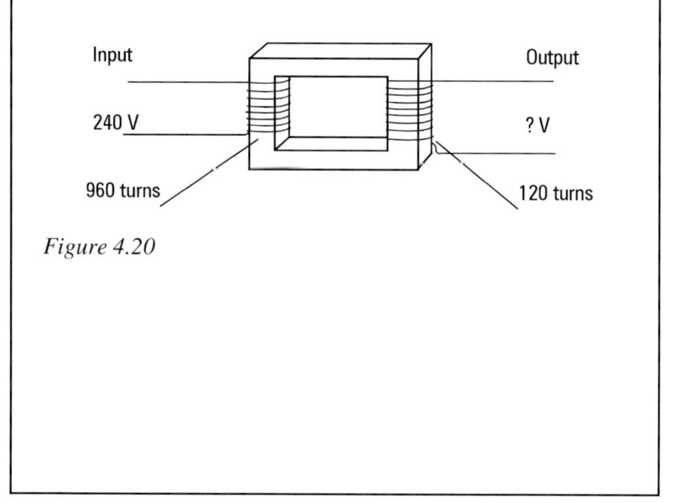


*Figure 4.20*

## Current ratio

When we looked at transmission voltages we saw that the voltage was increased so that power could be distributed over long distances at reduced currents. This allowed the use of small conductors on pylons.

Transformers are the source of the increased transmission voltages and their output windings reflect the reduced current used.

If we ignore any inefficiency, the input power and the output power of a transformer will be the same. This means that if the voltage goes up the current comes down. Or, if the voltage is stepped down the current will increase.

### Example

The calculation of current in the windings of a transformer still uses the ratio of the number of turns in each coil.

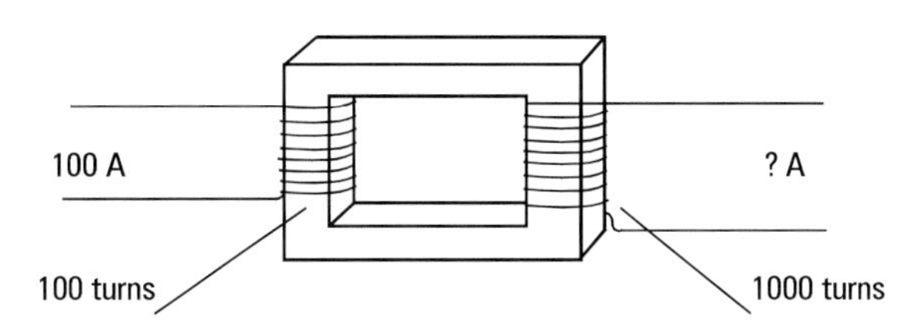


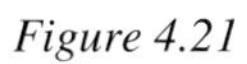
*Figure 4.21*

If we take the example in Figure 4.21, which has a turns ratio of

$$100:1000$$
$$\text{or} \quad 1:10$$

the current in the input is 100 A, so the current in the output is

$$= \frac{100}{10}$$
$$= 10 \text{ A}$$

### Example

A transformer (Figure 4.22) has a turns ratio of 25:1. The current on the primary side is 180 A. Calculate the output current.

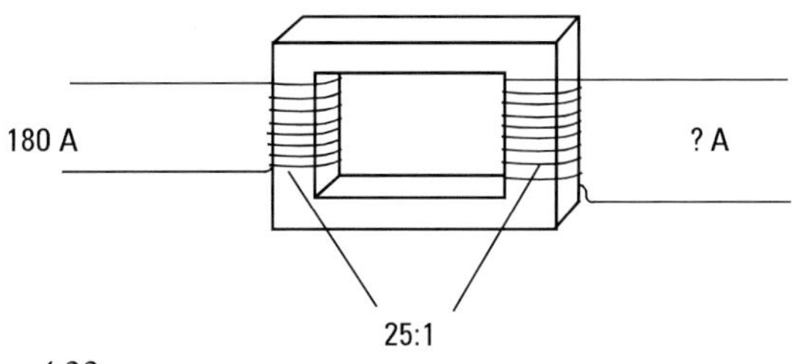


*Figure 4.22*

We know the output current will be greater than 180 A, so the 25:1 ratio must be used as a multiplying factor.

$$180 \times 25 = 4500 \text{ A}$$

### Try this

A transformer has a turns ratio of 8:1. The current on the low voltage side is 120 A. Calculate the current on the high voltage side.

## The iron core

Until now we have assumed that the iron core of the transformer is made of solid metal (Figure 4.23). If it is an electrical conductor as well as magnetic, an e.m.f. is induced in it at the same time as the second winding. This e.m.f. creates large currents for the iron core, which is a complete circuit.

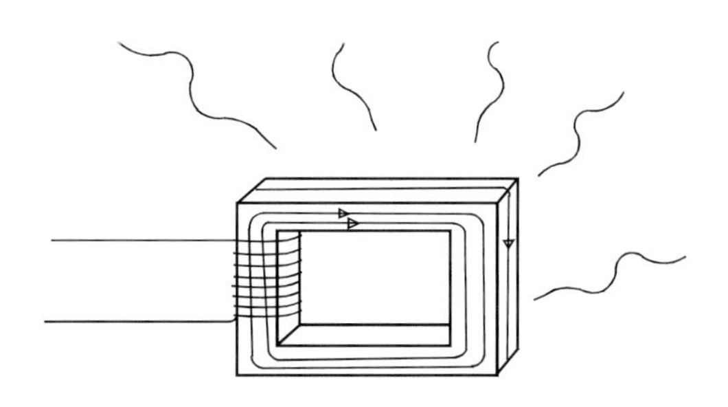

*Figure 4.23*

The ideal iron core is one which conducts the magnetic field well but does not conduct current. To achieve this the core is made up of very thin layers of silicon iron sheet, each layer insulated from the other (Figure 4.24).

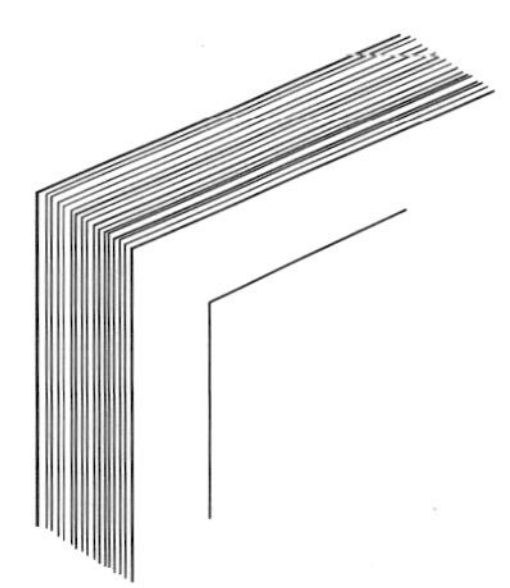

*Figure 4.24*

This laminated core cuts the electrical losses to a minimum but still allows it to be a very good magnetic conductor.

### Remember

**An iron core is not a solid object but is made up of many laminations, all insulated from each other.**

## Three-phase transformers

As we have seen, electricity is generated in large quantities using three phases. So that electricity can be transmitted and distributed over long distances, the voltages are stepped up using three-phase transformers. For high-voltage transmission purposes a three-wire delta connection (Figure 4.25) is used, but for lower voltage to the consumer a four-wire star winding (Figure 4.26) provides the output.

*Figure 4.25* *Delta-connected winding*

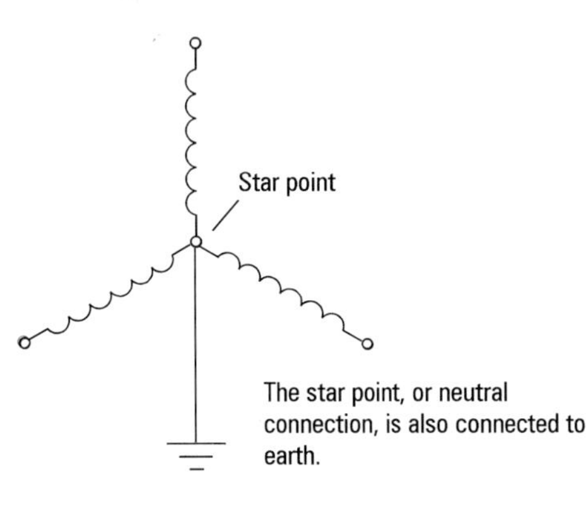


*Figure 4.26* *Star-connected winding*

## Voltage relationships

Whereas on a delta-connected winding the voltage across each section is the same, a star-connected winding has two voltages. If we consider the star winding output of a distribution transformer (Figure 4.27) the voltage across any two phases is 400 V, whereas between any phase and neutral it is 230 V.

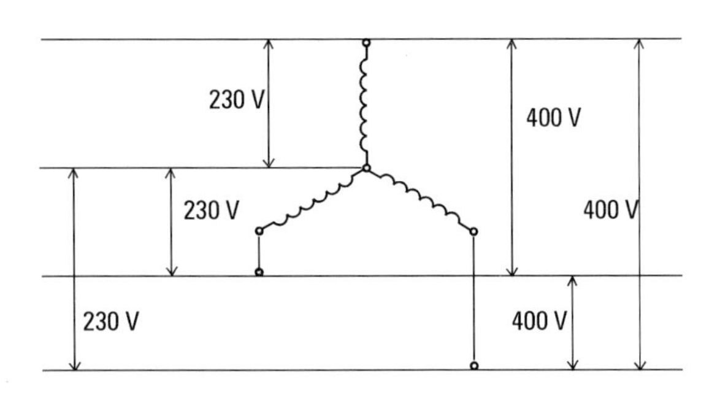


*Figure 4.27*

## Load balancing

So that a transformer is evenly balanced on each phase winding a number of steps are taken. Where electricity is being distributed in areas where there are a large number of single-phase loads an equal number of loads are connected to each phase (Figure 4.28).

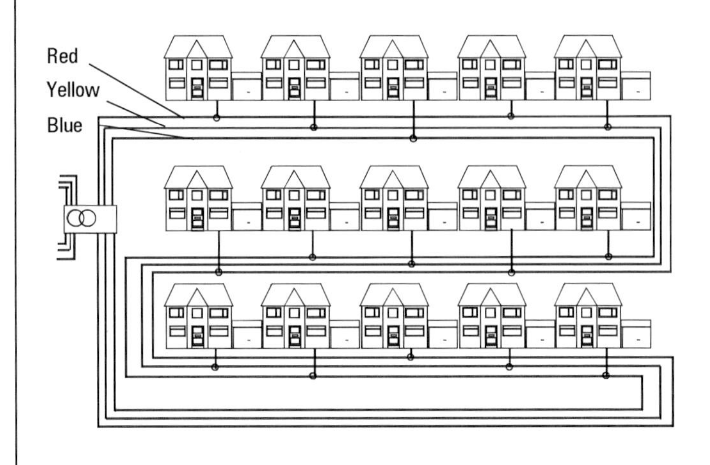


*Figure 4.28* *Houses are connected so that the load is spread across the three phases*

This same system is used throughout a three-phase installation so that all phases carry similar currents.

## Induction

As we can see, transformers work on the principle of induction from one coil to another. This same theory does, however, exist when there is only one coil and it is connected to an a.c. supply. Alternating current continually changes polarity, and this can be shown by considering a d.c. current first connected one way and then the other, as in Figures 4.29 and 4.30.

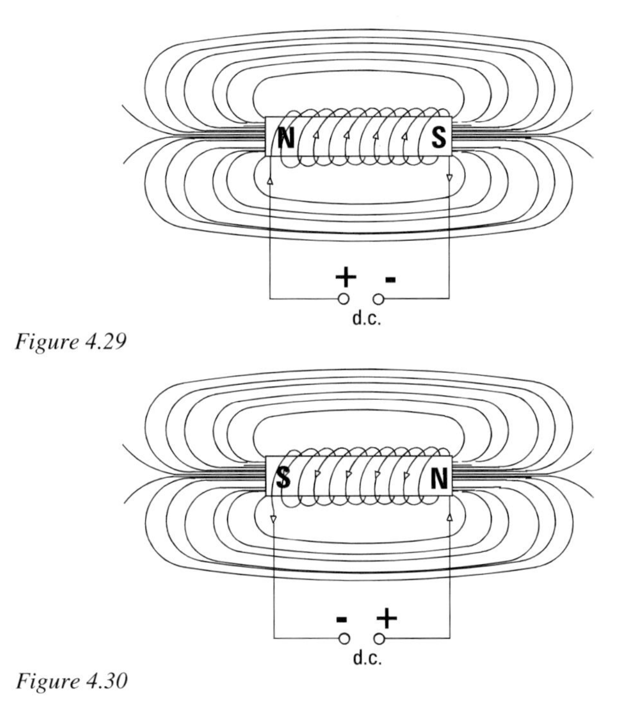


*Figure 4.29*

*Figure 4.30*

When it is connected in the first way, the magnetic field builds up with the North and South as shown. If it is now suddenly reversed the first magnetic field has to collapse and the new one build up in the opposite direction. This action takes time and can produce heat. The time has the strange effect of making the current flow in the circuit 90° after the voltage: Figure 4.31.

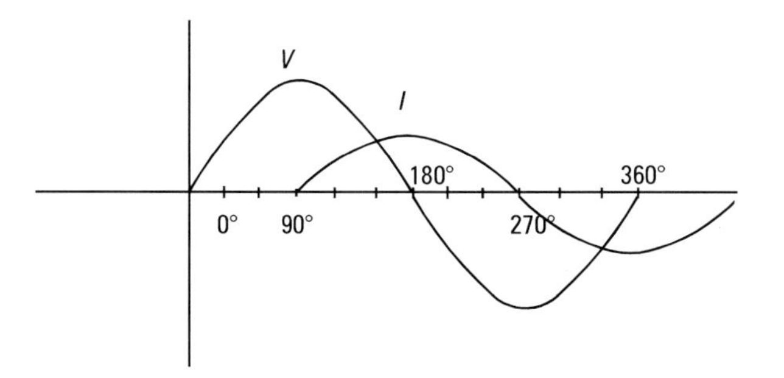


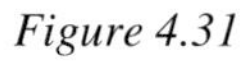
*Figure 4.31*

This of course causes complications when using

$$VI = P$$

for at no instant are the voltage and current at the same part of their waveforms at the same time.

In a.c. circuits anywhere where coils are used with iron cores a different formula has to be used to calculate power. This is

$$P = VI \cos \phi$$

where $\cos \phi$ is the cosine of the angle showing the difference between the voltage and current waveforms.

The $\cos \phi$ factor is referred to as the power factor. There are special meters available for measuring the phase angle between voltage and current in an a.c. circuit, known as power factor meters (Figure 4.32). Power factor meters are connected in a similar way to wattmeters, with separate current and voltage coils.

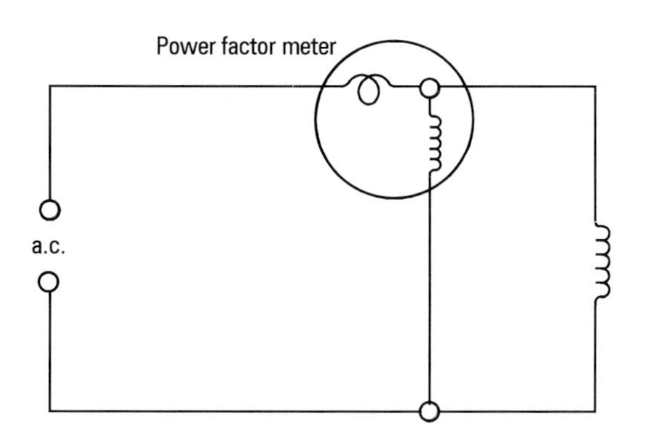


*Figure 4.32*

*Points to remember* ◀ - - - - - - - - - - - - - - - -

## Transformers

A double-wound transformer consists essentially of ______ insulated from each other and mounted on ________________ ____________

The output voltage of a step-up transformer will be ________ than the input voltage.

A step-down transformer has 80 turns on the output side. If we need to step down the voltage from 240 V to 12 V how many turns must be on the input side?

If the input power and the output power of a transformer is the same, what happens to the current if the voltage is stepped up?

How is the iron core of a transformer constructed?

Electricity is generated in large quantities using three phases. For high-voltage transmission a ______________________ connection is used, but for lower voltages a ________________ provides the output.

The output between two phases of a distribution transformer is ____ V, but between phase and neutral it is ____ V.

The $\cos \phi$ factor is often referred to as ________________

$\cos \phi$ is the cosine of the angle showing the difference between the ____________ and current waveforms.

# Part 3

## D.C. supplies

Most of the direct current (d.c.) we use is from cells and batteries, but d.c. can be generated similarly to a.c. When generating single-phase a.c. a coil is rotated in a magnetic field and a sine wave results. To convert that output to d.c. the negative part of the waveform is reversed so that both halves are positive. To achieve this a rotating switch called a **commutator** is used instead of the slip rings (Figures 4.33 and 4.34).

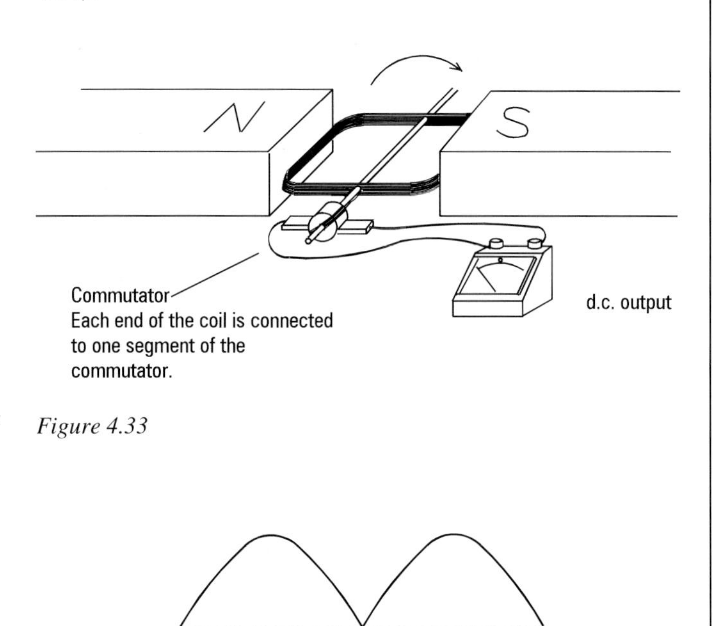


*Figure 4.33*

*Figure 4.34* *Full-wave d.c. waveform*

The effect of converting a.c. into d.c. can also be carried out without moving parts by using diodes and rectifiers. A diode is a device which will only allow current to pass through it in one direction. When the current is reversed the diode will not conduct. By placing a single diode in a single-phase circuit, all of one half of the waveform can be cut off (Figure 4.35).

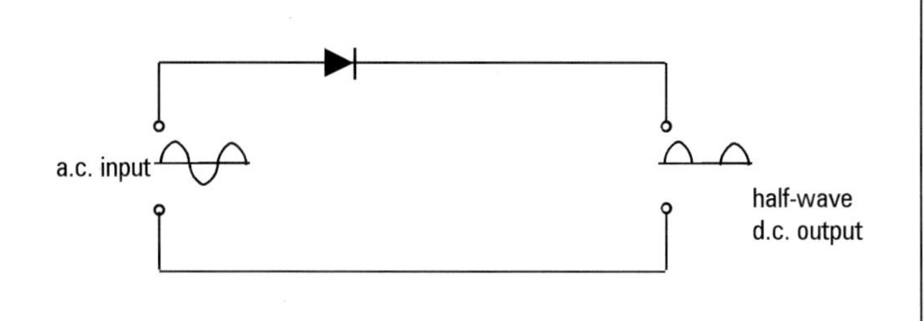


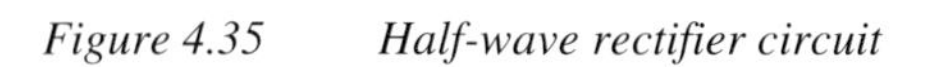

*Figure 4.35* *Half-wave rectifier circuit*

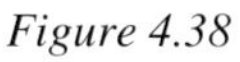

To create a "full-wave" d.c. similar to the output of the generator four diodes are required to be connected, as shown in Figure 4.36.

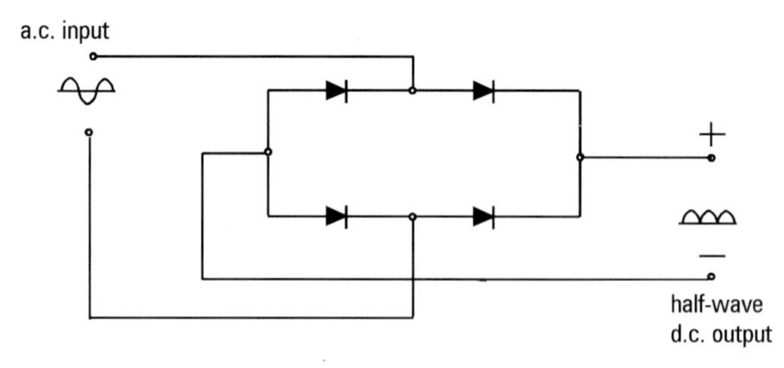


*Figure 4.36*

In a full-wave bridge rectifier all diodes point towards the positive of the d.c.

The bridge rectifier is often displayed as a diamond as shown in Figure 4.37.

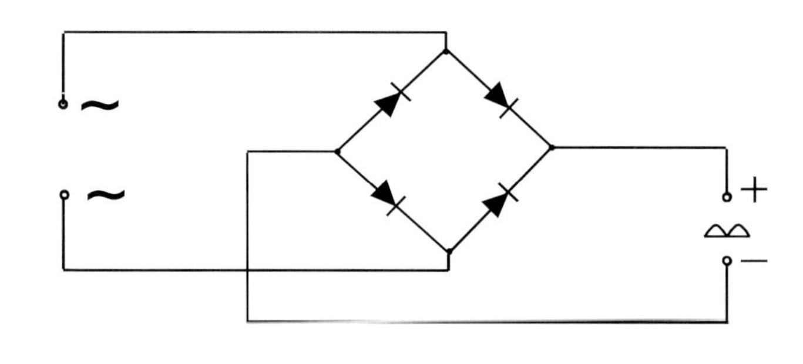


*Figure 4.37*

~ is the symbol for a.c. supply.

Where a centre-tapped transformer is used, a full-wave circuit can be devised using two diodes, as shown in Figure 4.38.

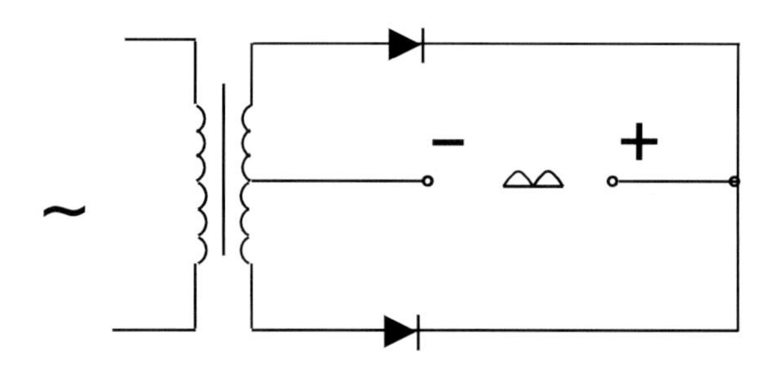


*Figure 4.38*

## Production of an e.m.f. by chemical means

A cell is a store of electrical energy in a chemical form. When a cell is connected to a circuit, chemical action inside the cell produces a voltage to make things work (Figure 4.39).

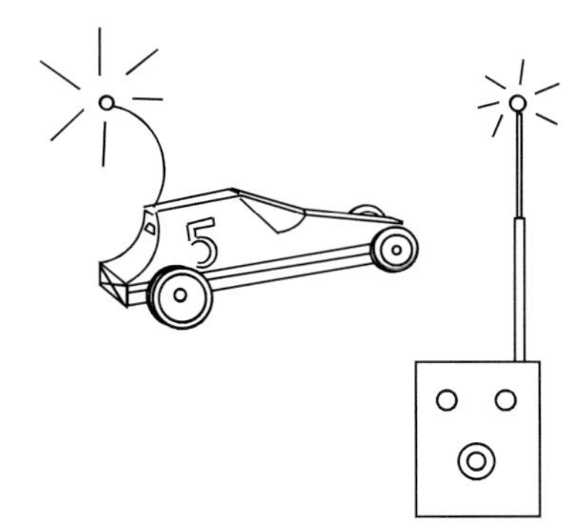


*Figure 4.39 Radio-controlled car*

After a time the chemical action deteriorates and the voltage drops below a useful level. Eventually the chemical action stops and the cell is of no further use.

## Cell construction

To construct a simple cell three things are needed: two different materials or plates – one for the positive, the other for the negative – and an electrolyte. The electrolyte causes a chemical reaction to start when the two plates are placed in it. A simple cell of this type (Figure 4.40) can be made using copper and zinc as the two plates and a liquid electrolyte containing sulphuric acid and water.

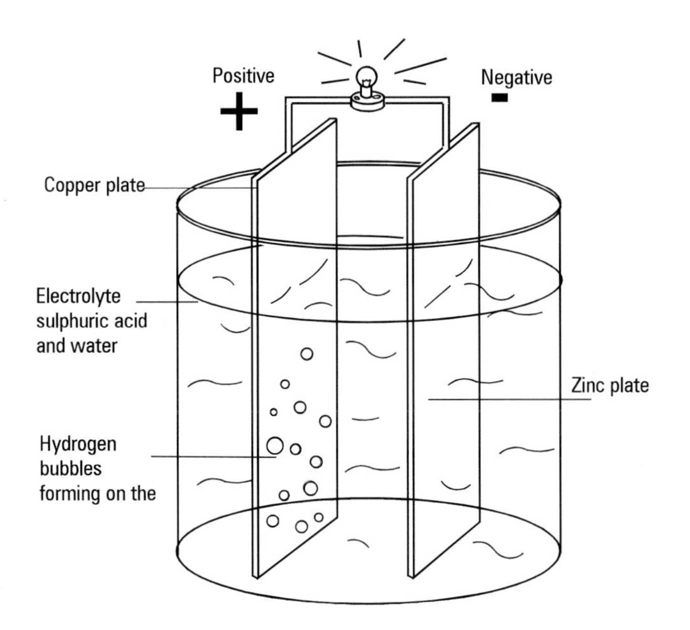


*Figure 4.40 A simple or Voltaic cell*

Cells can be made from different materials to give various voltages. Here are some examples.

*Table 4.1*

| Cell type | Positive material | Negative material | Electrolytes | Voltage produced |
|---|---|---|---|---|
| Voltaic | Copper | Zinc | Dilute sulphuric acid | 0.5–1.0 V |
| Law | Carbon | Zinc | Ammonium chloride | 0.5–1.0 V |
| Leclanché | Carbon | Zinc | Ammonium chloride | 1.4 V |
| Danielle | Copper | Zinc | Copper sulphate & zinc sulphate | 1.17 V |
| Weston Standard | Mercury | Cadmium | Cadmium sulphate | 1.083 V |
| Zinc carbon | Carbon | Zinc | Ammonium chloride & zinc | 1.5 V |
| Alkaline | Zinc powder | Manganese | Potassium hydroxide | 1.5 V |
| Mercury button | Zinc powder | Mercuric oxide | Alkaline | 1.5 V |

### *Try this*

Look around your home for all the different uses you can find for dry cells. Don't forget to look for door-bells, clocks and so on.

Investigate a portable radio, tape recorder or other equipment using several dry cell batteries.

Find out:
1. How many batteries are used?
2. What is the labelled voltage of each battery?
3. What voltage do they supply for the electrical equipment to use?

## Cells connected in series

To give a higher voltage cells are often connected in series. The voltage provided by each cell adds together to give a higher total voltage. For example, a 9 V battery contains six small 1.5 V cells connected in series in a casing.

Figure 4.41 shows the electrical symbol for a single cell as used in circuit diagrams.

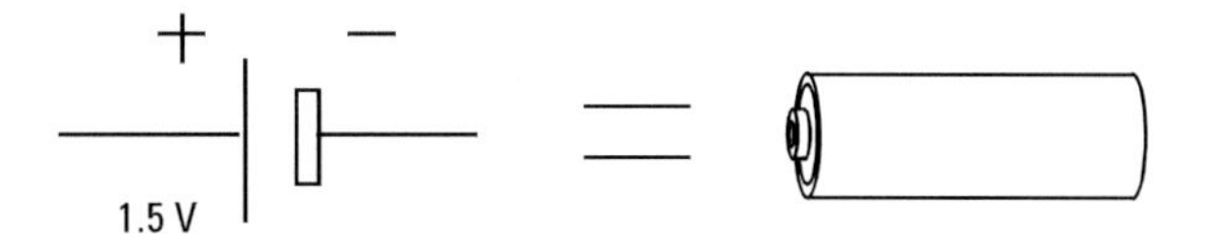


*Figure 4.41*

When two cells are connected in series the positive pole of one cell is connected to the negative pole of the other cell (Figure 4.42).

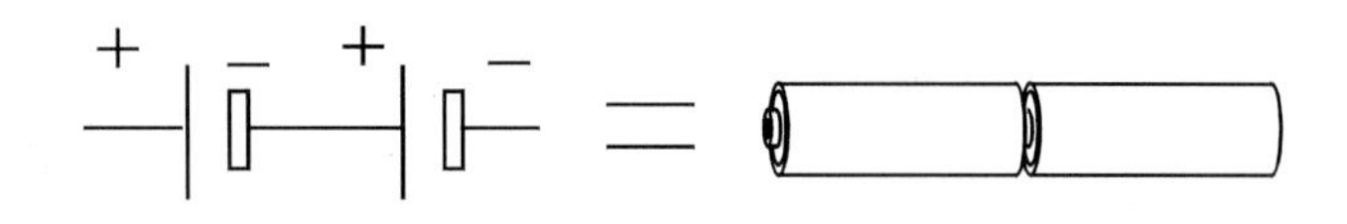


*Figure 4.42*

Cells connected like this are called batteries, although this name is also commonly used for single cells. A car battery is in fact a set of cells linked together in series inside a case to produce 12 volts. Each cell can produce about 2 volts; therefore six cells are used.

The electrical symbol would look like that shown in Figure 4.43.

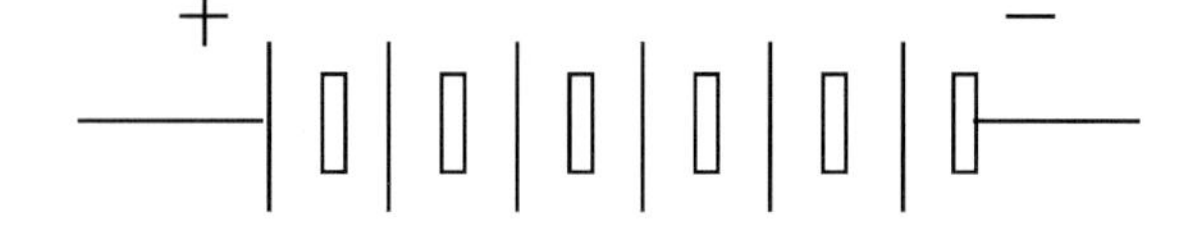


*Figure 4.43*

But usually, to make it easier, any multicell battery has the symbol shown in Figure 4.44.

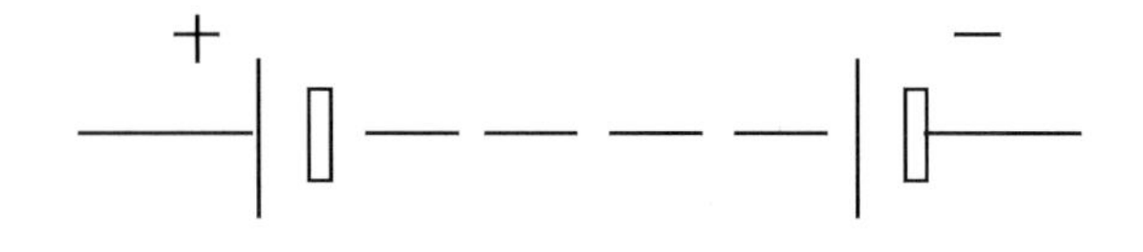


*Figure 4.44*

## Cells connected in parallel

Cells can be connected in parallel; that is, the two positive poles are joined and the two negative poles are joined.

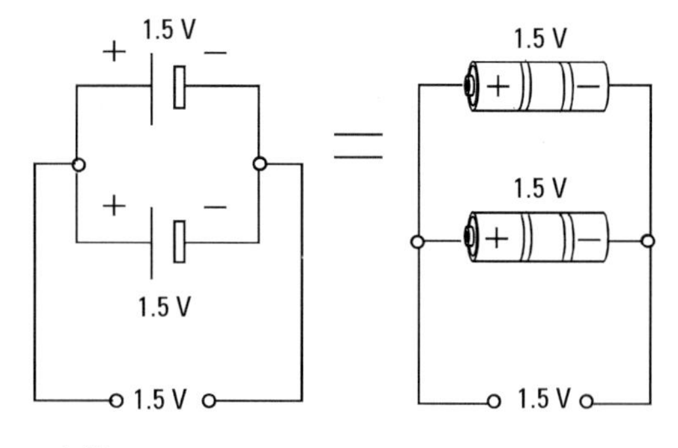


*Figure 4.45*

As you can see from Figure 4.45, the parallel connection of cells does not increase or add the voltage, it stays the same. But cells connected like this can provide the same power for twice as long; they share the load. In effect this makes a "bigger" battery of the same voltage. This system is used where a long life is important.

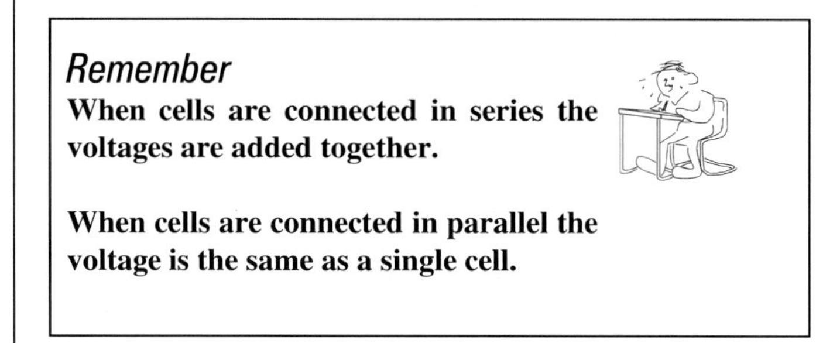

### *Remember*

**When cells are connected in series the voltages are added together.**

**When cells are connected in parallel the voltage is the same as a single cell.**

## Secondary cells

Primary and secondary cells can be very similar to look at, but they are in fact different.

A primary cell produces electricity from a chemical reaction taking place in the cell when it is connected to a circuit; a secondary cell uses chemical action, but in effect it stores electricity in a chemical form. When the chemical action of a primary cell slows down and stops the cell is exhausted and must be replaced. A secondary cell, however, can be recharged with electricity many times over, giving it a very long life!

The chemical storage of electrical energy in a primary cell can be likened to the heat energy stored in a match head. Once the energy is released it cannot be replaced. A secondary cell may, in the same comparison, be considered as a fuel-filled lighter, where once the energy is released the fuel can be renewed and the lighter used over again.

So secondary cells store electricity in a chemical form, that is, a chemical change takes place to produce electricity. When the battery is run down it can be recharged from an electrical supply. Recharging reverses the chemical process in the battery. This discharging and recharging can be repeated many times.

## The lead acid cell

These are the most common type of secondary cells. As the name implies, the cell is constructed from lead for the active plates and an acid electrolyte. The anode or negative material is made from spongy lead. The cathode or positive material is made from lead dioxide (lead peroxide), and the electrolyte is dilute sulphuric acid. The latest lead acid cells use a gel and not a liquid for the electrolyte and need not therefore only be used in an upright position.

## Construction

Each cell has a number of positive and negative plates or grids which carry the active lead materials. The plates in a cell are always an odd number, having one extra negative plate. The positive plates are interleaved with the negative plates, so a 13 plate cell will have six positive and seven negative plates.

## The battery

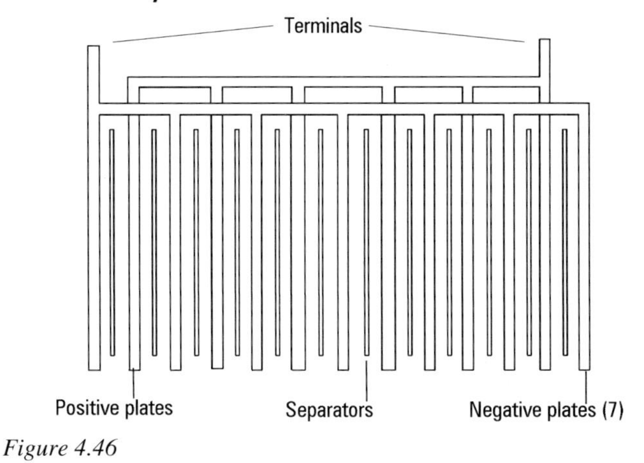


*Figure 4.46*

A lead acid battery (Figure 4.46) is made up from a set of cells fitted together in a single casing. Each cell produces about 2 volts and as the cells are joined or connected in series a 12 volt car battery will have six cells (Figure 4.47).

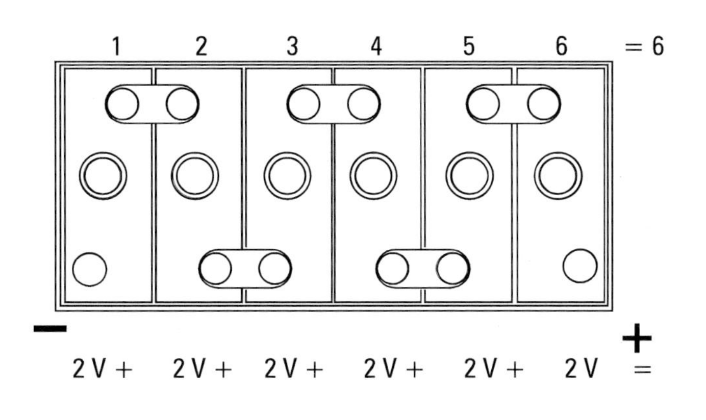


*Figure 4.47*

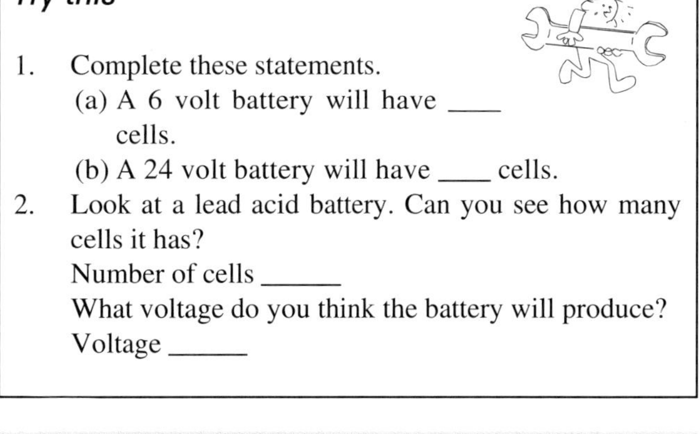

### *Try this*

1. Complete these statements.
   (a) A 6 volt battery will have ____ cells.
   (b) A 24 volt battery will have ____ cells.
2. Look at a lead acid battery. Can you see how many cells it has?
   Number of cells ______
   What voltage do you think the battery will produce?
   Voltage ______

**Battery electrolyte is very corrosive – avoid contact with the acid electrolyte!**
**Do not cause any electrical short circuits across the battery terminals.**

### *Points to remember*

Direct current can be generated in a similar manner to a.c. by converting the a.c. output using a rotating switch. The rotating switch is called a ______________

An a.c. output can also be achieved by using a full-wave bridge rectifier which requires _______ __________ to be connected.

Cells store electrical energy in a chemical form. In a simple cell you will find three different components. Name the three components in the most common secondary cell, the lead acid cell.

How do you obtain the total voltage when cells are connected in series?

and in parallel?

Draw the electrical symbol for a single cell.

# Part 4

## Battery charging

As we have seen, a secondary cell or battery can be recharged from an electrical supply source. However, this supply must be of the correct type, voltage and current to suit the battery.

**Mains voltage must never be connected directly to any battery!**

A special electrical device called a battery charger is used to recharge a battery. These come in many different shapes and sizes, but their main functions are the same:

1. Correct supply

The household or general workshop electrical mains supply is an alternating current (a.c.). This means that the electrical current is constantly changing its direction. Secondary cells can only be charged by current flowing in one direction, direct current (d.c.). The battery charger converts the alternating supply to direct current.

2. Correct voltage

Mains voltage is much too high (about 240 volts). The battery charger (Figure 4.48) reduces this to a safe voltage which is suitable for the type of battery being charged; for example, a 12 volt car battery requires about 14 to 16 volts charging voltage.

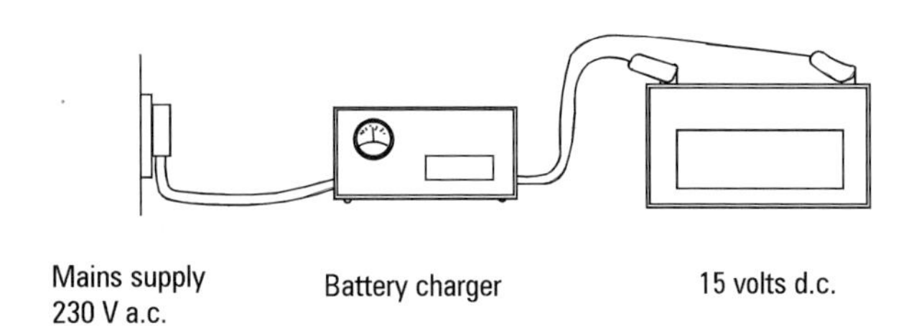


*Figure 4.48*

3. Correct current flow

The battery charger must also control the current flow to the battery. Too high a current will overheat and damage the battery!

## Using a battery charger

Make sure you have a battery charger which will supply the correct voltage and current for the battery. Most smaller battery chargers self-regulate the current at a fairly low level. They are often called trickle chargers. If you are charging a lead acid vehicle battery there are safety rules which you must observe:

1. Protect the circuits

Modern equipment and vehicles often have electronic components which can be damaged by sudden higher voltages which may be produced by the charger. To be safe disconnect the battery from what it is supplying before putting it on charge.

2. Correct polarity

The charger must be connected the correct way round (correct polarity) by connecting the positive lead to the positive terminal and the negative lead to the negative terminal. Battery terminals are marked + and –. On batteries with round post terminals the positive post is always larger in diameter.

3. Connecting up and switching off

A lead acid battery gives off hydrogen gas when being charged, the higher the charging current the more gas is produced. Hydrogen gas is highly flammable therefore great care must be taken when charging a battery.

**DON'T** disconnect or connect leads to the battery with the charger switched on

**DON'T** allow any electrical sparks or naked flames near the charging battery

**DON'T** let anyone smoke in the area

**DO** charge batteries in a well ventilated area

**DO** switch off the mains supply before connecting or disconnecting the charger leads

4. Electrolyte

The electrolyte is made from sulphuric acid ($H_2SO_4$) and water. This acid is corrosive and can cause personal injury, in particular to the eyes.

The acid is also harmful to components, paintwork and clothing, so avoid spillage and any contact with the skin and eyes or clothes. If contact occurs wash with plenty of clean water and if it is splashed in your eyes seek medical advice after washing.

## Testing and servicing

Batteries can be tested to measure their state of charge and to find out if the battery is still serviceable.

Testing can be done in several ways:

1. Battery voltage

With a voltmeter connected (Figure 4.49), a 12 volt lead acid vehicle battery in good condition should read about 12.5 volts. If the battery has just been charged the voltage will be higher, about 13–13.5 volts. If the voltage is much below 12 volts the battery may need attention.

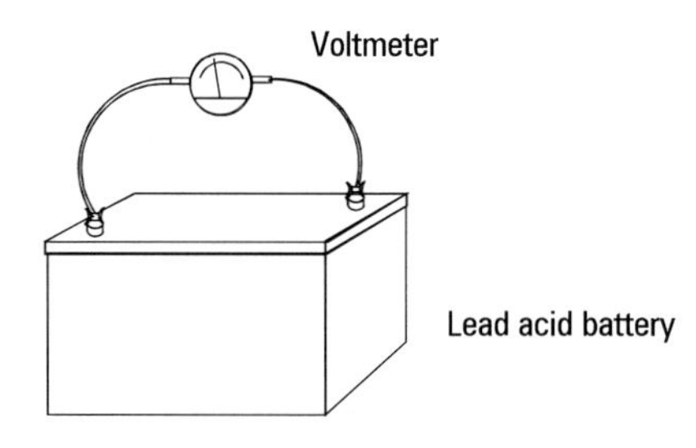


*Figure 4.49 Voltage test*

2. Electrolyte specific gravity

The specific gravity or density of the electrolyte changes depending on the state of charge from mostly water when discharged to full strength when fully charged. This can be measured using an instrument called a hydrometer (Figure 4.50).

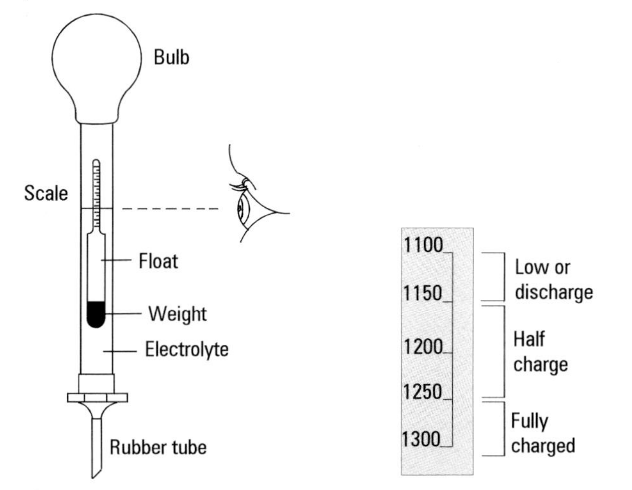


*Figure 4.50 The hydrometer and scale*

By depressing the rubber bulb electrolyte is drawn up into the hydrometer. When the float has settled the scale reading level with the electrolyte can be read.

3. A high rate discharge tester

This has heavy duty connecting leads, a voltmeter scale and a very large circuit load which can be switched on and off (Figure 4.51).

When the tester is connected to a battery it will read the battery voltage.

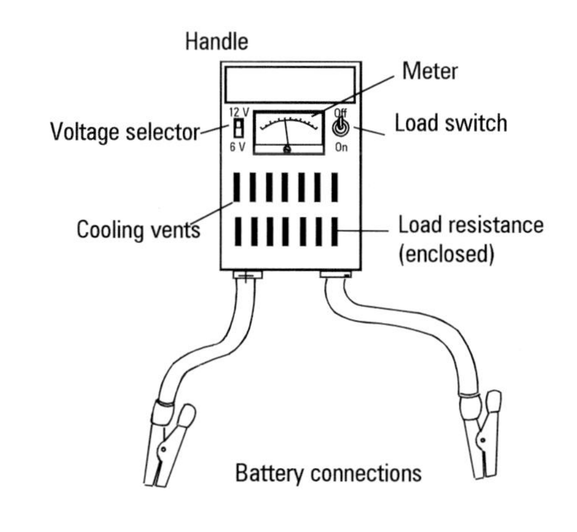

*Figure 4.51 Load tester*

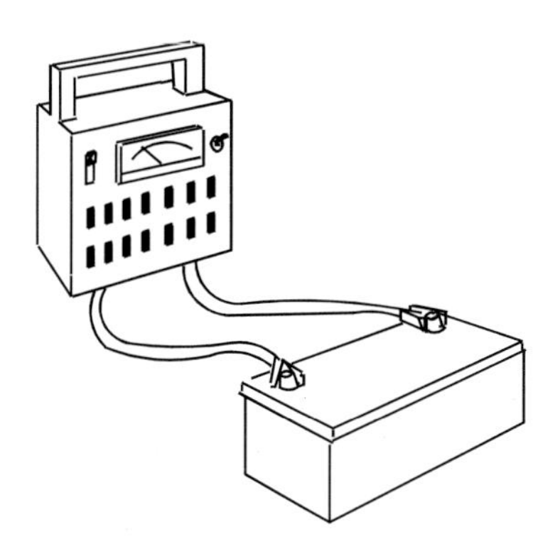

*Figure 4.52 A load test (drop test) being carried out on a lead acid cell.*

The load resistance is switched on (Figure 4.52), and this imposes a very heavy load on the battery; in fact it is an equivalent load to using a vehicle starter motor.

The current flow under load is very heavy (perhaps 100–200 amperes); therefore the test can only be applied for a short time, about 10 seconds, or the battery may be damaged.

The load can be felt in the quantity of heat flowing from the vents.

The battery voltage is checked before the test, under load conditions and after testing. The battery must be in good charge condition before the test: 12.5 volts or above. If it is not then it must be charged up first.

Under load the battery voltage should not fall below 10.0 volts and should return after the test to about 12 volts or above.

## Servicing

With all batteries it is important to keep the battery posts and terminals free from corrosion and scale. They should be tight and lightly coated with grease or petroleum jelly to prevent corrosion (Figure 4.53). Some batteries are maintenance-free and require no other servicing, but batteries with vent tops need to be checked for their electrolytic level. When the battery is charging some of the electrolyte water evaporates or is lost by electrolysis, so the level should be topped up with distilled water to keep out dirt and other deposits which could damage the battery. The correct level may be marked on the battery, but if not the level should just cover the plates by about 5 mm. Any more than this will weaken the electrolyte.

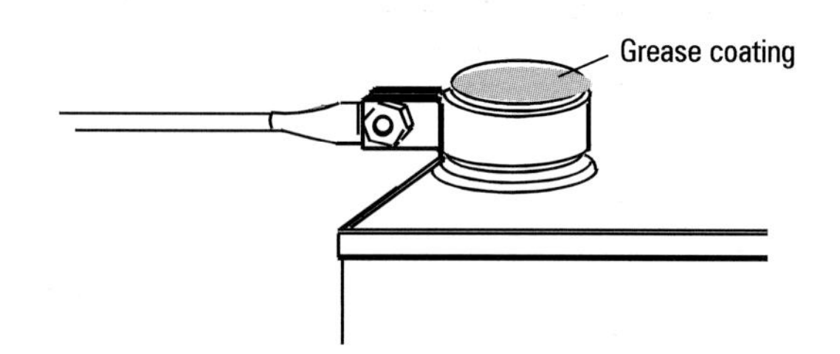


*Figure 4.53* *Battery terminal*

## Battery capacity

Battery capacity is a measure of the quantity of current or power which a battery can produce and for how long it can produce it, with the voltage staying above 10.8 volts (for a 12 volt battery). Capacity depends on the area of the plates and the number of plates in each cell; this means that a large battery will give the same voltage, but will give a higher current or a similar current flow for a longer time.

Battery capacity can be calculated on the 10 hour rate or the 20 hour rate. A battery which produces 5 amperes for 10 hours equals 50 ($5 \times 10$) ampere hours at the 10 hour rate. The same battery could give 3 amperes for 20 hours, so would equal 60 ampere hours at the 20 hour rate.

In recent years many developments have taken place in the design of new types of cells. The chemicals used as electrolytes can be very different and are not interchangeable. Care must therefore be taken to use the correct chemicals when maintaining batteries.

### *Points to remember*

Secondary cells or batteries can be recharged from an electrical supply source by a battery charger. The battery charger converts the alternating supply to direct current and reduces mains voltage to a charging voltage of approximately ____ volts.

When charging a lead acid vehicle battery safety rules which must be observed include:

A hydrometer measures the specific gravity of the ________

## *Self-assessment multi-choice questions*

**Circle the correct answers in the grid below.**

1. When a magnet is pushed into a coil and then left stationary in the coil the meter would show
   (a) a deflection in one direction and then return to zero
   (b) a deflection in one direction and then stay there
   (c) a deflection first in one direction and then the other
   (d) no deflection at all
2. The output from a rotating coil of a single-phase alternator is by
   (a) permanent magnet
   (b) slip rings
   (c) springs and conductors
   (d) continuous connection by wire
3. If the voltages between the phases of a three-phase star-connected supply is 400 V, the phase to neutral voltage will be approximately
   (a) 692 V
   (b) 240 V
   (c) 230 V
   (d) 2150 V
4. The effect of having two or more cells connected in series will be to
   (a) add the cell voltages
   (b) divide the cell voltages
   (c) last twice as long
   (d) remain the same
5. In Figure 4.54 the cells are connected
   (a) x series and y series
   (b) x parallel and y series
   (c) x series and y parallel
   (d) x parallel and y parallel

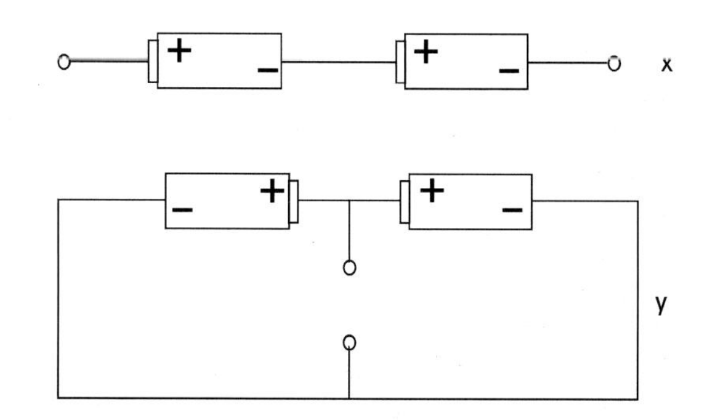


*Figure 4.54*

### *Answer grid*

| | | | | |
|---|---|---|---|---|
| 1 | a | b | c | d |
| 2 | a | b | c | d |
| 3 | a | b | c | d |
| 4 | a | b | c | d |
| 5 | a | b | c | d |

# Progress Check

**Circle the correct answers in the grid at the end of the multi-choice questions.**

1. The volume of a rectangular prism 0.5 m × 0.5 m × 2 m is
   (a) 0.5 $m^3$
   (b) 2 $m^3$
   (c) 3 $m^3$
   (d) 50 $m^3$
2. The input power of a motor is 8 kW. If the output power is 5 kW, what is the percentage efficiency of the motor?
   (a) 40%
   (b) 62.5%
   (c) 80%
   (d) 87.5%
3. The prefix milli is used to denote a submultiple of
   (a) $10^{-12}$
   (b) $10^{-9}$
   (c) $10^{-6}$
   (d) $10^{-3}$
4. Mass is measured in
   (a) kilograms
   (b) newtons
   (c) joules
   (d) coulombs
5. The SI symbol for density is ρ and it is measured in
   (a) $m/s^2$
   (b) Nm
   (c) $kg/m^3$
   (d) Wb
6. What is the work done when a load of 5 kg is lifted 0.2 m from the ground?
   (a) 245.25 joules
   (b) 1.00 joules
   (c) 9.81 joules
   (d) 39.24 joules
7. What effort would be required to lift a load of 10 kg placed 600 mm from the fulcrum of a lever? The distance between the fulcrum and the point the force is to be exerted is 750 mm.
   (a) 8 N
   (b) 78.48 N
   (c) 1.23 N
   (d) 784.8 N
8. Quantity of electrical charge is measured in
   (a) amperes
   (b) coulombs
   (c) watts
   (d) volts
9. An electrical appliance connected to a 230 V supply and taking 13 A consumes a power of
   (a) 17.69 kW
   (b) 38.87 kW
   (c) 2.99 kW
   (d) 56.52 kW
10. How long would a current of 5 amperes have to flow for if 1500 coulombs of electricity are used?
    (a) 5 minutes
    (b) 10 minutes
    (c) 15 minutes
    (d) 20 minutes
11. Given that the total voltage in a circuit is 12 V and the total current is 5 A then the resistance is
    (a) 300 Ω
    (b) 60 Ω
    (c) 2.4 Ω
    (d) 0.417 Ω
12. A circuit supplied with d.c. has a current rated at 2 A and a resistor in the circuit rated at 10 Ω. Calculate the power in the circuit.
    (a) 5 W
    (b) 12 W
    (c) 20 W
    (d) 40 W
13. How many cycles per second is electricity generated at in the UK and Europe on a public mains supply?
    (a) 50
    (b) 100
    (c) 150
    (d) 200
14. A transformer has 600 turns on the input winding and 150 turns on the output winding. If the input voltage is 240 V what will the output be?
    (a) 960 V
    (b) 480 V
    (c) 120 V
    (d) 60 V

15. Calculate the current in the output of a transformer with windings as shown in Figure 4.55.
    (a) 12.5 A
    (b) 25 A
    (c) 100 A
    (d) 200 A

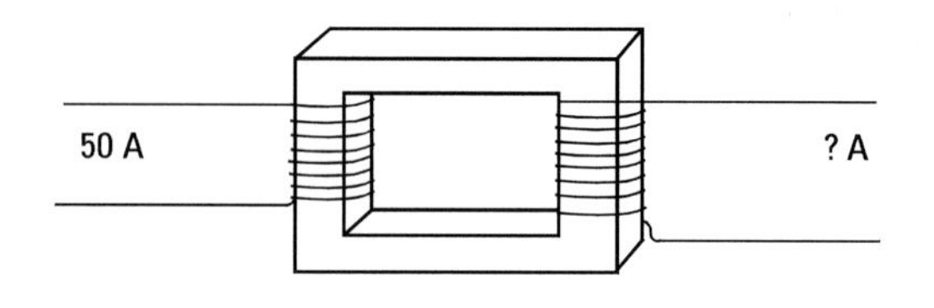


*Figure 4.55*

16. Inductance is measured in
    (a) farads
    (b) henrys
    (c) coulombs
    (d) ohms
17. A "commutator" is
    (a) a rotating switch
    (b) slip rings
    (c) a waveform
    (d) a type of meter
18. In a single phase circuit how many diodes are required to be connected to create "full-wave" d.c. similar to the output of a generator?
    (a) 1
    (b) 3
    (c) 4
    (d) 8
19. To recharge a 12 V lead acid cell the charging voltage would be
    (a) 240 V d.c.
    (b) 15 V d.c.
    (c) 10 V d.c.
    (d) 6 V d.c.
20. The core of a transformer is made from
    (a) silicon steel laminations
    (b) aluminium strips
    (c) shaped copper
    (d) solid iron
21. The star point of a transformer winding is connected to
    (a) the delta point and earth
    (b) two phases and earth
    (c) red phase and neutral
    (d) earth and neutral
22. If the voltage between the phases of a three-phase star connected supply is 440 V, the phase to neutral voltage will be approximately
    (a) 761 V
    (b) 440 V
    (c) 254 V
    (d) 240 V
23. A hydrometer measures the
    (a) specific gravity
    (b) voltage
    (c) current
    (d) electrolyte level
24. A hydrometer reading for a fully charged lead acid battery is
    (a) 1100
    (b) 1750
    (c) 1200
    (d) 1250
25. When being charged a lead acid battery gives off
    (a) oxygen
    (b) water
    (c) lead acid gas
    (d) hydrogen gas

## Answer grid

| | | | | | | | | | |
|---|---|---|---|---|---|---|---|---|---|
| 1 | a | b | c | d | 16 | a | b | c | d |
| 2 | a | b | c | d | 17 | a | b | c | d |
| 3 | a | b | c | d | 18 | a | b | c | d |
| 4 | a | b | c | d | 19 | a | b | c | d |
| 5 | a | b | c | d | 20 | a | b | c | d |
| 6 | a | b | c | d | 21 | a | b | c | d |
| 7 | a | b | c | d | 22 | a | b | c | d |
| 8 | a | b | c | d | 23 | a | b | c | d |
| 9 | a | b | c | d | 24 | a | b | c | d |
| 10 | a | b | c | d | 25 | a | b | c | d |
| 11 | a | b | c | d | | | | | |
| 12 | a | b | c | d | | | | | |
| 13 | a | b | c | d | | | | | |
| 14 | a | b | c | d | | | | | |
| 15 | a | b | c | d | | | | | |

# 5

# Introduction to Electronics

Complete the following statements to remind yourself of some important facts from the previous chapter.

The piezoelectric effect converts __________ energy into electrical energy.

To produce an e.m.f. by magnetic means requires two main components, ______________ and ________________

A cell is a store of electrical energy in a ___________ form.

What are the two differences we noted between primary and secondary cells?

**On completion of this chapter you should be able to:**

- carry out basic calculations for voltage, current and resistance
- use component suppliers' catalogues for finding out component data
- carry out calculations to determine the values of equivalent resistors
- use the resistance colour and number codes to determine the values of resistors
- recognise and use the tolerance values of resistors
- relate the value of capacitance to plate size, thickness and type of dielectric
- recognise the basic construction of some capacitor types
- identify the polarity of a polarised capacitor
- relate the connection of capacitors in series or parallel to form equivalent values of capacitance
- compare the construction of low-frequency with high-frequency inductors
- compare inductors with transformers
- calculate voltage and turns ratio of double-wound transformers
- recognise the meaning of "p" and "n" semiconductor materials
- identify the output of a circuit which consists of a single diode which is supplied with a.c.
- recognise the need to connect an LED to the correct polarity
- identify some of the basic semiconductor devices
- recognise some of the basic graphical symbols for electronic components
- complete the revision exercise at the beginning of the next chapter

## Part 1

Electricians can no longer justify saying that they have no need to learn about electronics, for the use of devices such as light dimmers, boiler controllers and speed controllers has now brought electronics very much into general electrical installation work.

The fact that most electronic equipment works on milliamperes means that care has to be taken by the electrician who is used to dealing with much larger currents. If a circuit is not checked first then a test with a 500 volt insulation tester can write off many pounds worth of electronic equipment.

Although unit replacement is often used rather than component replacement, a basic knowledge of components can be very helpful. It is not important at this stage to have a great depth of knowledge of electronic components. It is, however, important to have some basic understanding of

- how they work or their action
- what they are used for
- their basic construction
- how to identify them

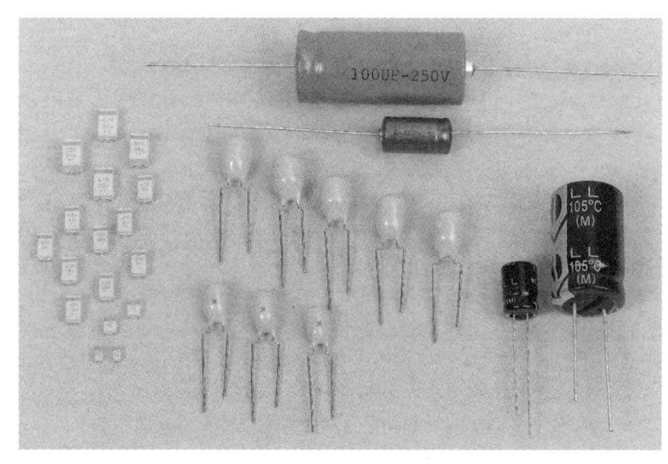


*Figure 5.1*

**Safety**
**When carrying out work with electronic components and circuits there are a number of safety factors that must be taken into account.**

Soldering, by its very nature, requires high temperatures to make it work. Careless use of equipment can cause burns, and in extreme cases, fire. The safety of the user and others must always be given serious consideration.

The Electricity at Work Regulations 1989 make it very clear that you should never work on or near live equipment unless every possible precaution has been taken to prevent injury. It is necessary on occasions to take readings with instruments to measure voltage or current. Wherever possible the meter should be connected when the supply is off. The supply can then be switched on for the readings to be taken. If readings have to be taken with a supply connected, appropriate shields, screens and test equipment must be used. Even after the supply has been disconnected it is possible to get a nasty shock from capacitors that have remained charged.

You should also remember that your body can hold very high static charges of electricity, and should this become discharged through some electronic components it will destroy them. Often when these components are supplied by the manufacturer they are in special antistatic containers. Care must always be taken when handling these components and special precautions should be taken.

One other precaution to note is that some electronic components may contain beryllium oxide and should not be cut open or incinerated. Always follow manufacturers' instructions.

## Identification of components

One of the main aims of this part of the book is to help you to recognise and identify the different components used in electronics. With all of the components being miniaturized, identification is becoming more difficult. In the past each type of component had its own characteristics, but with modern encapsulation methods these are less clear. However, this book concentrates on the common components and deals with possible ways of recognising them (Figure 5.2).

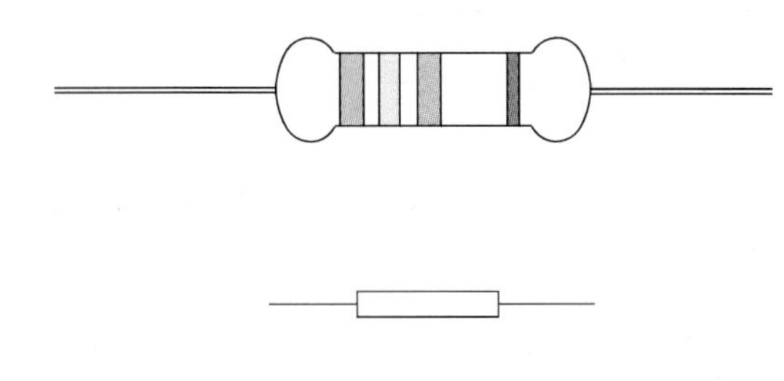

*Figure 5.2* *Colour-coded resistor and general symbol*

# Resistors

Resistors are one of the most common electronic components there are.

They are connected in circuits for several different reasons, but generally to control voltages and/or currents.

When determining what resistance is required, or what effect a resistor has, a formula derived from Ohm's Law can be used.

$$R = \frac{V}{I}$$

$$V = IR$$

$$I = \frac{V}{R}$$

## Types and uses of resistors

Resistors can be used for voltage control (potential dividers). Within a circuit it is often necessary to have different voltages at different stages. This can be achieved by using resistors and creating voltage drops.

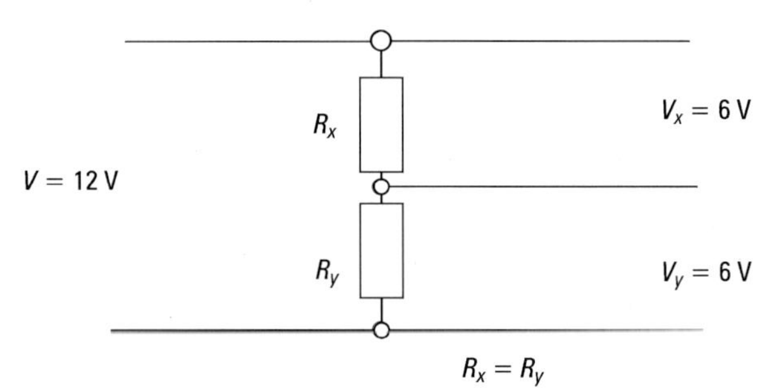


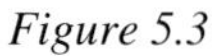

*Figure 5.3*

In the circuit shown in Figure 5.3, when resistors *x* and *y* have exactly the same value then the voltage will be divided equally between them.

If the voltages were to be 1 V and 11 V (Figure 5.4) then the resistors would be in the ratio of 1:11. For example, resistor values of 10 Ω and 110 Ω may have been selected depending on the actual current requirements.

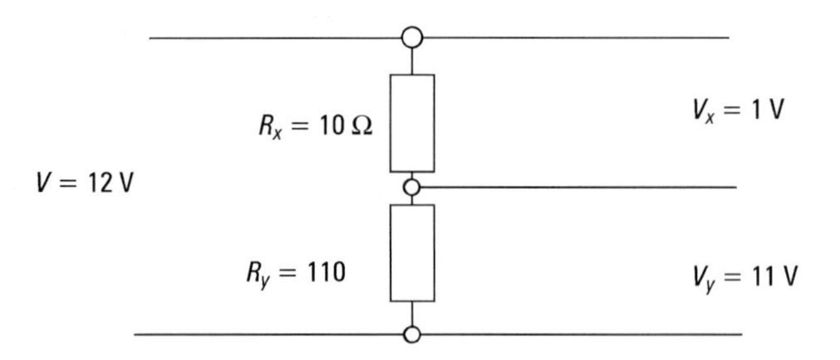


*Figure 5.4*

$$\frac{R_x}{R_y} = \frac{V_x}{V_y}$$

### *Example*

In the circuit in Figure 5.5 $R_x$ has a value of 100 Ω, voltage $V_x$ is 3 volts and voltage $V_y$ is 9 volts. Calculate the value of $R_y$.

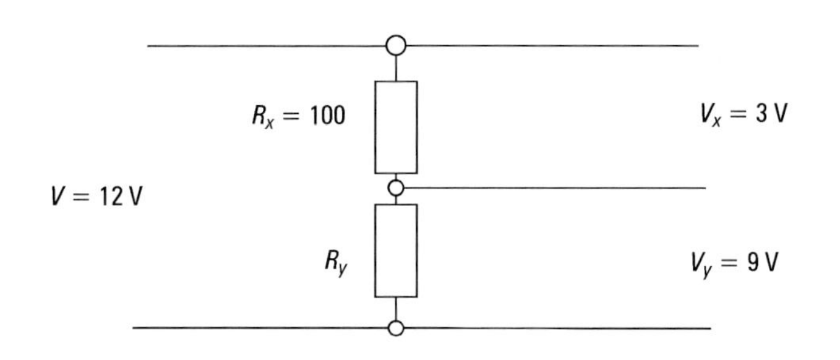


*Figure 5.5*

$$\frac{R_x}{R_y} = \frac{V_x}{V_y}$$

$$\frac{100}{R_y} = \frac{3}{9}$$

$$R_y = 300\ \Omega$$

### *Try this*

In the circuit shown in Figure 5.6 calculate the value of $R_a$.

2200 Ω

V = 230 V

$R_a$

10 V

*Figure 5.6*

$R_a =$

## Resistors used for controlling current (shunts)

Where shunt resistors are used to control currents they have to be connected into the flow of current, in other words in parallel with the load.

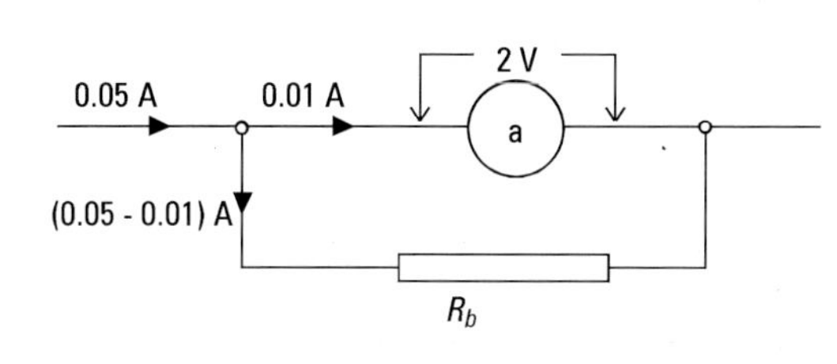

*Figure 5.7*

In Figure 5.7 the circuit current is 0.05 A (50 mA), but the component indicated as "a" can only cope with a current of 0.01 A. $R_b$ is connected across it to "shunt" the current past. The voltage across the component "a" is 2 V, so the same voltage will be across $R_b$.

The current through $R_b$ is calculated by:

$$0.05 - 0.01 = 0.04\text{ A}$$

The value of $R_b$ can now be calculated:

$$R_b = \frac{V}{I} = \frac{2}{0.04} = 50\ \Omega$$

If the voltage is increased the shunt resistor will not offer protection.

## Power rating

Resistors often have to carry comparatively large values of current, so they must be capable of doing this without overheating and causing damage. As the current has to be related to the voltage it is the power rating of the resistor that must be identified.

Power is calculated by the voltage multiplied by the current.

$$P = V \times I$$

Instead of $V$ in the equation we can substitute $I \times R$ (from Ohm's Law), or instead of $I$ we can substitute $V/R$.

So we can now also calculate power from:

$$P = I^2 R \quad \text{or} \quad P = \frac{V^2}{R}$$

The power rating of the 50 Ω resistor calculated for Figure 5.7 would be

$$P = VI = 2 \times 0.04 \qquad = 0.08\text{ watts}$$

or

$$P = I^2 R = 0.04^2 \times 50 \qquad = 0.08\text{ watts}$$

or $$P = \frac{V^2}{R} = \frac{2^2}{50} = 0.08 \text{ watts}$$

Normally only one calculation is necessary.

Typical maximum power ratings for resistors are shown in Figures 5.8–5.10.

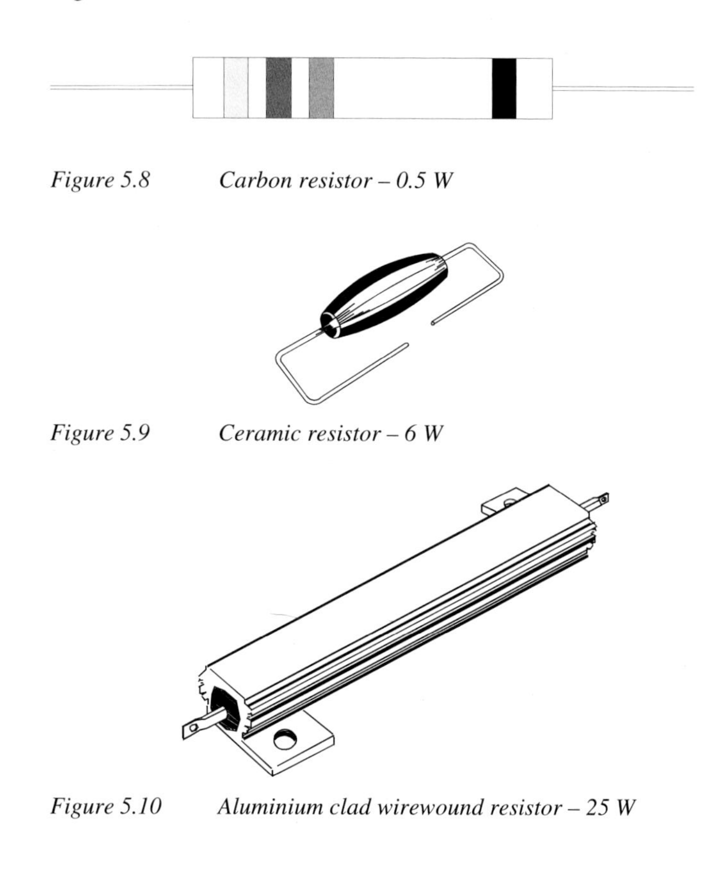

*Figure 5.8* *Carbon resistor – 0.5 W*

*Figure 5.9* *Ceramic resistor – 6 W*

*Figure 5.10* *Aluminium clad wirewound resistor – 25 W*

The construction of these and other resistors are covered in some detail further on in this section.

### *Try this*

Look at an electronics components catalogue and list the power ratings for each of the following types of resistors.

| | Typical power ratings |
|---|---|
| Carbon | |
| Wire-wound ceramic | |
| Wire-wound aluminium clad | |

# Equivalent resistance

## Series circuit

Unfortunately it is not always possible to have the exact single resistor with the correct combination of resistance and power required. In these circumstances a number of resistors can be connected together.

If a greater value of resistance is required a number of single resistors can be connected in series.

Total resistance =
resistance 1 + resistance 2 + resistance 3 + resistance 4 ... etc.

$$R_T = R_1 + R_2 + R_3 + R_4 \ldots$$

The power of each resistor can be calculated separately once the voltage or current is also known.

### *Try this*

In the following circuits (Figures 5.11 and 5.12), each of the resistors shown has to be replaced. As the exact single replacement resistor is unavailable each resistor must be replaced with at least two in series to give equivalent resistance and power. Use realistic values taken from a supplier's catalogue.

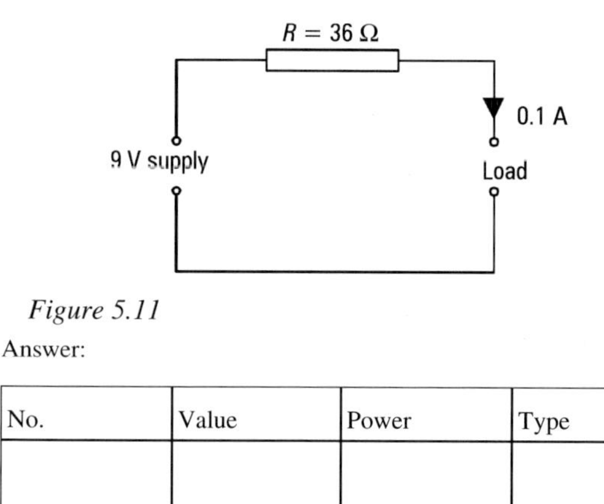


*Figure 5.11*

Answer:

| No. | Value | Power | Type |
|---|---|---|---|
| | | | |

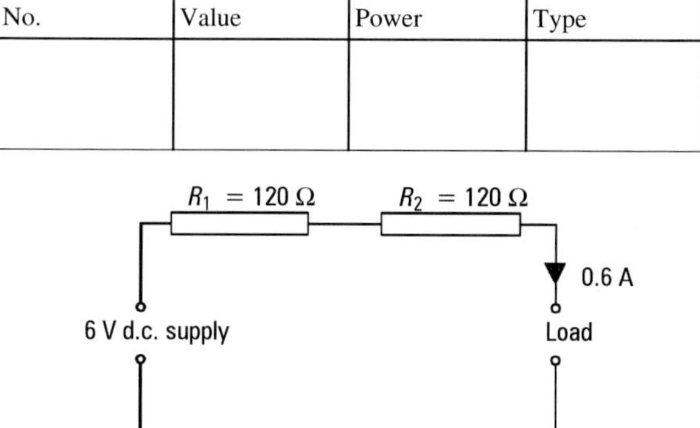


*Figure 5.12*

Answer:

| No. | Value | Power | Type |
|---|---|---|---|
| | | | |

> ### *Remember*
> 
>
> **When connecting two or more resistors in series to replace one, add their individual resistances to make the total resistance.**

## Parallel circuit

Calculating equivalent resistors for parallel or shunted circuits can be far more complex. Taking the situation calculated in the previous parallel circuit, shown in Figure 5.7, a resistor of 50 ohms carrying a current of 0.04 A was required. This one resistor could be replaced with two.

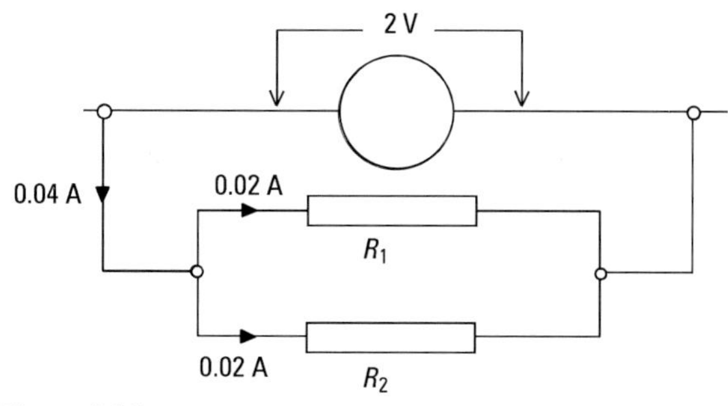


*Figure 5.13*

Two resistors of the same value could be used so that the current would be divided equally between them (as shown in Figure 5.13), but what value of resistance would be required? When two identical resistors are connected in parallel the combined total resistance is equal to half of one of the resistors. So, for example, if we used two 5 ohm resistors in parallel we would have a total of 2.5 ohms.

The formula for calculating the total resistance of a parallel circuit is

$$\frac{1}{\text{total resistance}} = \frac{1}{\text{resistance one}} + \frac{1}{\text{resistance two}}$$

$$\frac{1}{R_T} = \frac{1}{R_1} + \frac{1}{R_2}$$

or

$$R_T = \frac{R_1 R_2}{R_1 + R_2}$$

Assuming 5 Ω each

$$\frac{1}{R_T} = \frac{1}{5} + \frac{1}{5}$$

(use a calculator and divide 1 by 5 in each case)

$$\frac{1}{R_T} = 0.2 + 0.2$$

$$\frac{1}{R_T} = 0.4$$

$$R_T = \frac{1}{0.4}$$

(use a calculator and divide 1 by 0.4)

$$R_T = 2.5\ \Omega$$

This means that to end up with a value of 50 Ω, each resistor in Figure 5.13 will have to be 100 Ω. As the current through each has been reduced to 0.02 A the power rating will be $P = VI = 2 \times 0.02 = 0.04$ W (instead of 0.08 W).

Where there are more than two resistors in a parallel circuit the formula becomes

$$\frac{1}{R_T} = \frac{1}{R_1} + \frac{1}{R_2} + \frac{1}{R_3} + \frac{1}{R_4} \ldots \text{ etc.}$$

### *Example*

Calculate the values of the three identical resistors in parallel if the total resistance for the circuit in Figure 5.14 has to be 11 Ω.

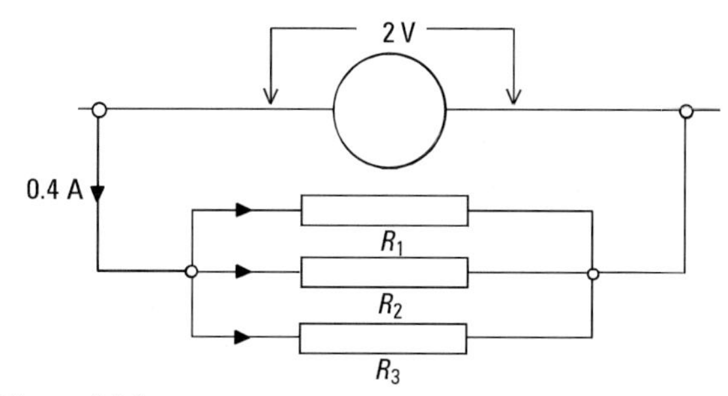


*Figure 5.14*

$$\frac{1}{R_T} = \frac{1}{R_1} + \frac{1}{R_2} + \frac{1}{R_3}$$

$$\frac{1}{11} = \frac{1}{R_1} + \frac{1}{R_2} + \frac{1}{R_3}$$

As $R_1 = R_2 = R_3$ we can call them all $R$

$$\frac{1}{11} = \frac{3}{R}$$

$$R = 33\ \Omega$$

So the value of each of the three resistors will be 33 Ω.

> ### *Try this*
> 
>
> In the circuit shown in Figure 5.15 the defective resistor shown has to be replaced with at least two in parallel to give equivalent resistance and power. Use realistic values taken from a supplier's catalogue.
>
> 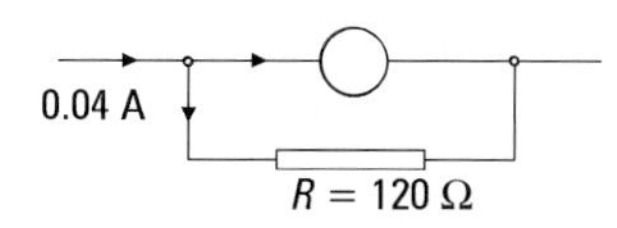
> 
>
> 
>
> *Figure 5.15*
>
> Answer:

| No. | Value | Power | Type |
| --- | --- | --- | --- |
| | | | |

## Remember

**When connecting two resistors in parallel to replace one, each resistor has to have twice the resistance, but half the power rating if they are identical.**

## Variable resistors

A large variety of variable resistors (Figure 5.16) are to be found in electronic circuitry. Their function varies from straightforward control to trimming and adjustment during commission and service routines.

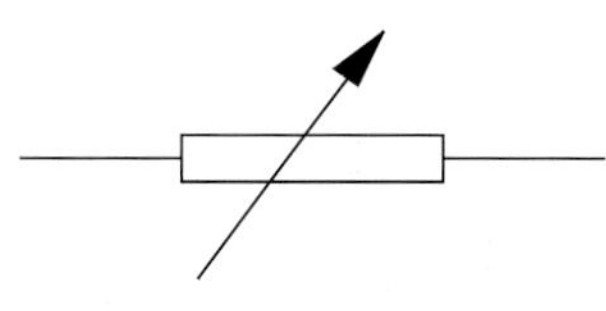

*Figure 5.16* *General symbol for variable resistor*

The variable resistor (Figures 5.17–5.19) is normally a three terminal device with connections to either end of a carbon track or wire-wound resistor and a third terminal connected to a sliding contact. In normal industrial practice the change in resistance is proportional to the movement of the slider: this is known as a linear track. In some cases, audio volume controls for example, the track is logarithmic and the change in resistance is not equal to the movement of the slider.

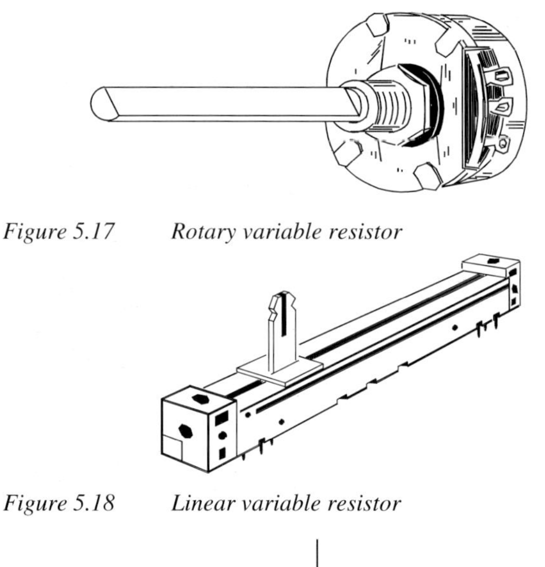

*Figure 5.17* *Rotary variable resistor*

*Figure 5.18* *Linear variable resistor*

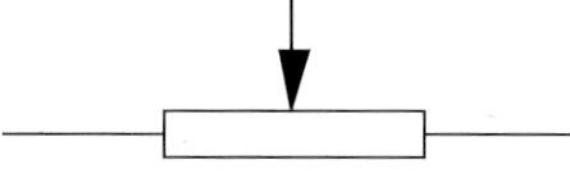

*Figure 5.19* *Symbol for voltage divider with moving contact (potentiometer)*

The resistance value quoted on the body of the variable resistor is that of the track, measured from end to end and the power rating of the device will be that given by the manufacturer's data. The variable resistor may also be referred to as a potentiometer or "pot" and may be listed in suppliers' catalogues under this description.

Standard panel-mounted potentiometers are normally designed to fit into a 9.5 mm hole with solder tags to give wired connection to the rest of the circuit.

Skeleton pots (Figure 5.20)and multiturn cermet trimmers (Figure 5.21) usually have pins designed for mounting directly on to circuit boards and are often used for pre-set adjustments (Figure 5.22) by the manufacturer.

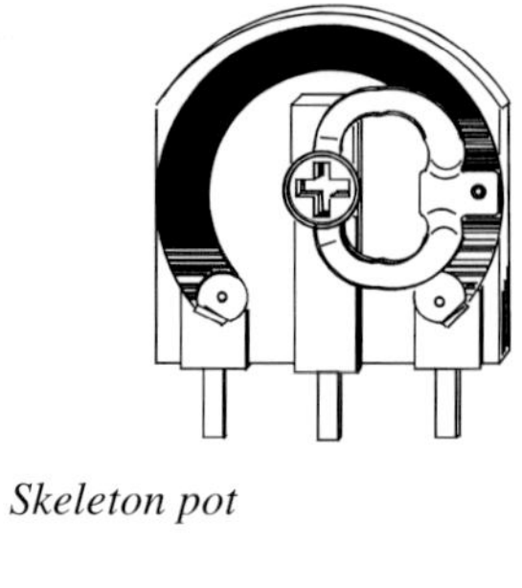

*Figure 5.20* *Skeleton pot*

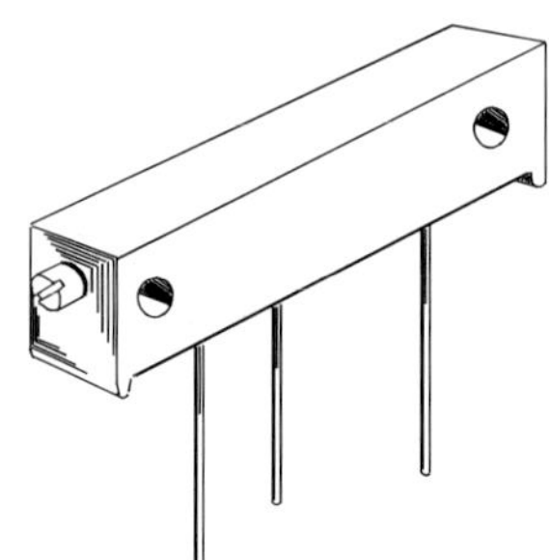

*Figure 5.21* *Multiturn cermet trimmer*

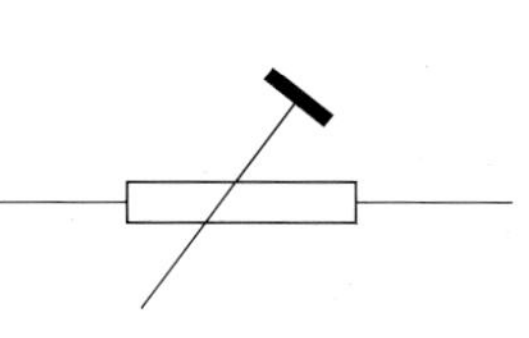

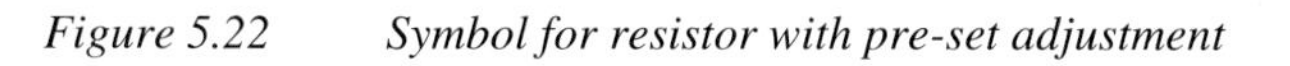

*Figure 5.22* *Symbol for resistor with pre-set adjustment*

## Try this

Use a supplier's catalogue to give the ordering details for the following.

All are variable resistors/potentiometers.

| Value | Type | Power | Catalogue number |
|---|---|---|---|
| 5 kΩ LIN | Carbon track | 2 W | |
| 220 Ω | Cermet | 5 W | |
| 250 Ω | Wire-wound | 1 W | |
| 2 kΩ | Multiturn | 0.25 W | |
| 50 kΩ | Trimmer | 0.75 W | |

## Thermistors

Thermistors are temperature sensitive resistors and may look similar to carbon resistors. However, it is far more likely they will take one of their other forms. Examples of some of these are shown in Figures 5.23–5.25.

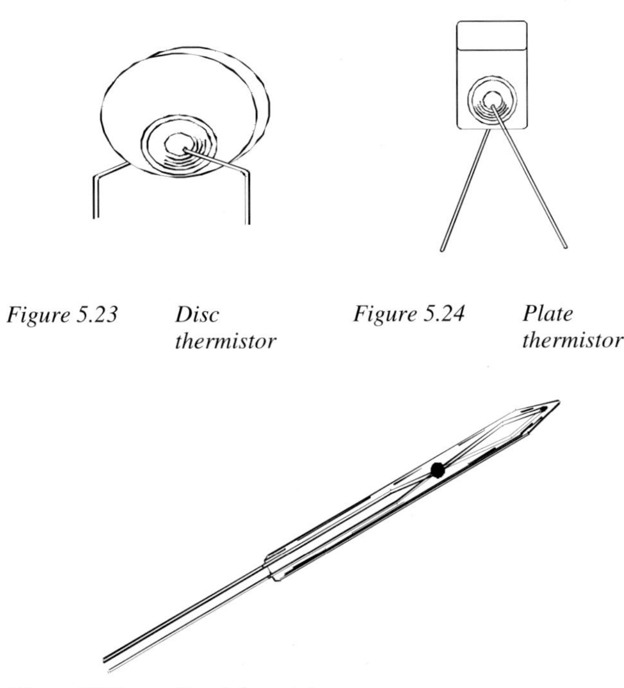

*Figure 5.23* *Disc thermistor*

*Figure 5.24* *Plate thermistor*

*Figure 5.25* *Bead thermistor*

Thermistors belong to a group of resistor made from semiconductor materials and are thermally sensitive. They have a controlled temperature coefficient which may be positive (PTC) or negative (NTC). Their uses include temperature measurement and control, temperature stabilisation, current surge suppression and a wide variety of other applications. They are suitable for both a.c. or d.c. circuits as they are not reactive or polarised.

The rated resistance of a thermistor may be indicated by the standard colour code or by a single body colour used just for thermistors. Typical values are shown in Table 5.1.

*Table 5.1*

| Resistance at 25 °C | Colour |
|---|---|
| 3 kΩ | red |
| 5 kΩ | orange |
| 10 kΩ | yellow |
| 30 kΩ | green |
| 100 kΩ | violet |

## Voltage-dependent resistors

These resistors are generally known as "varistors" and are another type of semiconductor resistor. They are principally used as voltage surge suppressors.

## Light-dependent resistors

In addition to temperature-sensitive resistors there are also light-sensitive resistors (Figure 5.26). These consist of a clear window with a cadmium sulphide film under it. When the light shines on the film its resistance varies. As the light increases the resistance reduces.

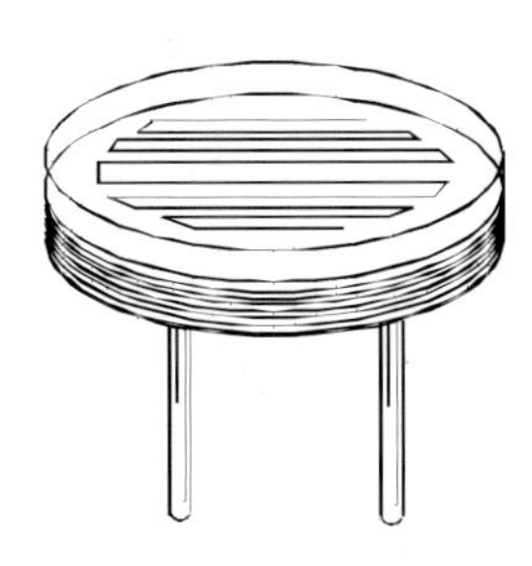

*Figure 5.26* *Light-dependent resistor*

## Resistor arrays

Resistor arrays are produced in encapsulated modules. These may be in a variety of configurations. Figures 5.27 and 5.28 show two possible units and how the resistors may be connected internally.

These are made for direct mounting onto printed circuit boards.

Pins number from pin 1 by the indicator in an anticlockwise direction

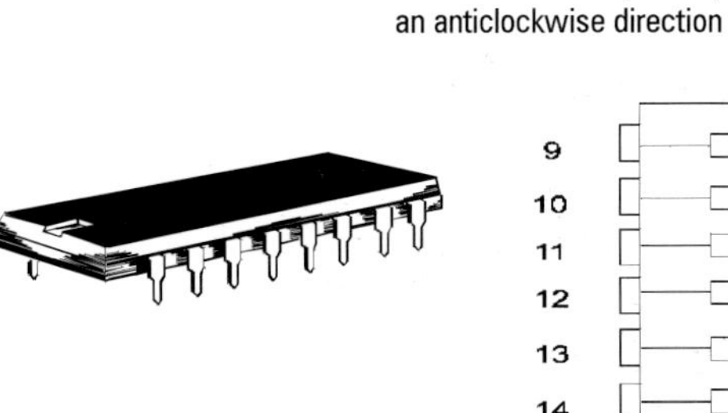

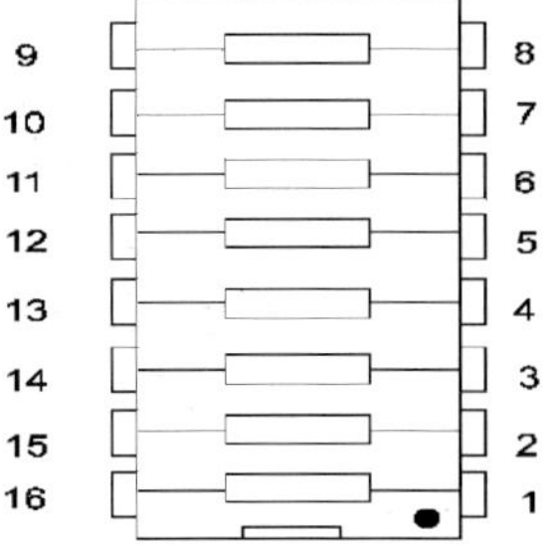


*Figure 5.27* *DIL (dual-in-line) 16 pins, 8 resistors*

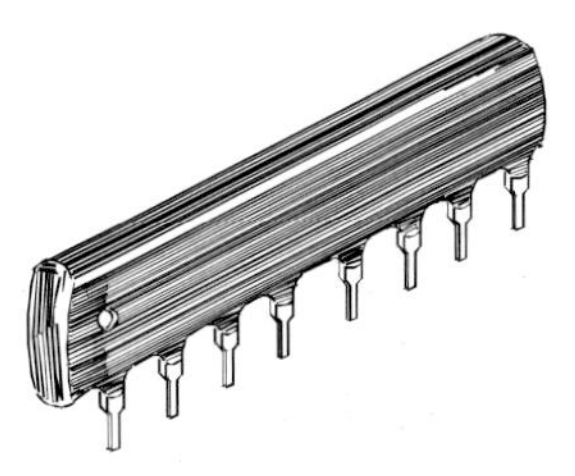

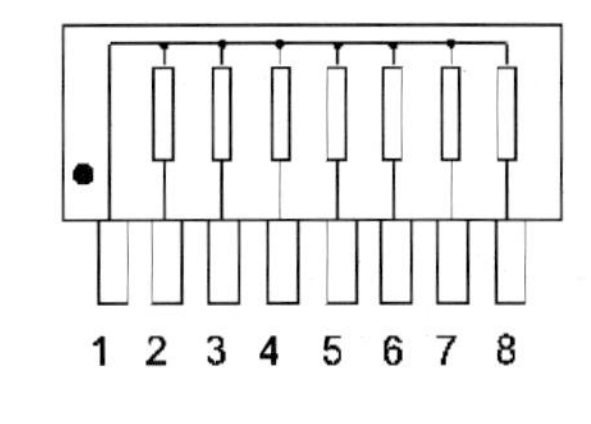


*Figure 5.28* *SIL (Single-in-line) 8 pins, 7 resistors*

## Points to remember

Resistors are one of the most common electronic components there are. They are generally connected in circuits to control ___________ and/or ___________

When resistors are used for voltage control they are called

_____________ _______________

When resistors are used for current control they are called _____________ _______________. In order to control the current the resistor has to be placed in ____________ with the load.

The three formulae for calculating power in a circuit are

When connecting two or more resistors in series to replace one their individual resistances must be _______ to make the total resistance.

When connecting two resistors in parallel to replace one each resistor has to have _______ the resistance, but _________ the power rating if they are identical.

A large variety of variable resistors are to be found in electronic circuitry for straightforward control and for trimming and adjustment purposes.

Thermistors are ______________ control resistors made from semiconductor materials. Their uses include temperature measurement and control.

Light-dependent resistors are sensitive to light. As the light increases, the resistance ____________

### Remember

Safety factors must be taken into account when carrying out work with electronic components and circuits.

The safety of the installer and others in the vicinity are of paramount importance.

Damage can also be caused to components through static charges of electricity held in the body.

*Always* take note of manufacturers' instructions.

## Self-assessment multi-choice questions

**Circle the correct answers in the grid below.**

1. Two resistors, one of 100 Ω and the other 200 Ω, are connected together in series. Which of the following series combinations would be suitable as a replacement?
   (a) 150 Ω, 100 Ω, 200 Ω
   (b) 30 Ω, 40 Ω, 50 Ω
   (c) 25 Ω, 150 Ω, 250 Ω
   (d) 150 Ω, 120 Ω, 30 Ω
2. The value of the resistor in Figure 5.29 would be

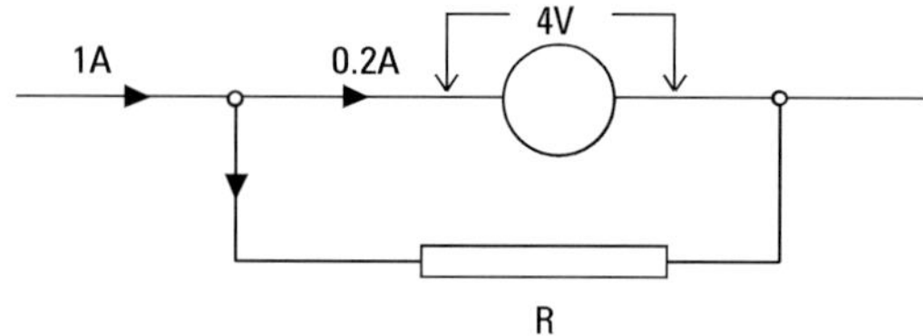


*Figure 5.29*

   (a) 5 Ω
   (b) 20 Ω
   (c) 4 Ω
   (d) 0.8 Ω

3. The graphical symbol in Figure 5.30 represents a

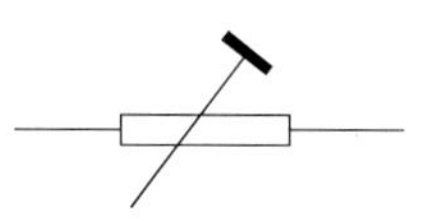

*Figure 5.30*

   (a) variable resistor with pre-set adjustment
   (b) voltage divider
   (c) potentiometer
   (d) general symbol for a resistor
4. When two identical resistors are connected in parallel the combined total resistance is equal to
   (a) the sum of the two resistors
   (b) the difference between the two resistors
   (c) half the value of one of the resistors
   (d) the value of one of the resistors
5. What would be the power rating of a 120 Ω resistor in a circuit supplied with 12 V?
   (a) 10 watts
   (b) 0.1 watts
   (c) 1.2 watts
   (d) 17 280 watts

### Answer grid

| | | | | |
|---|---|---|---|---|
| 1 | a | b | c | d |
| 2 | a | b | c | d |
| 3 | a | b | c | d |
| 4 | a | b | c | d |
| 5 | a | b | c | d |

# Part 2

## Construction of resistors

The construction of resistors falls into three groups:

- carbon
- film
- wire-wound

### Carbon

Carbon composition resistors have been in use since the earliest days of radio, but are now being replaced by film resistors. The carbon resistors, as they have always been known, have been made in power ranges from 0.1 watts to 2 watts and resistance values ranging from 10 ohms to 100 Megohms. They are still used as general-purpose resistors where temperature coefficient and close tolerance requirements are not important.

The carbon resistors consist of a resistance rod made up of fine carbon particles mixed with a non-conductive refractory filling which is bonded together by a resin binder (Figure 5.31). The value of resistance is determined by the proportion of carbon particles to filler. The end connections are moulded into the ends and then the complete unit is fired in a kiln. These may then be encapsulated in either a silicon lacquer or thermoplastic moulding, or placed in epoxy resin in a ceramic tube.

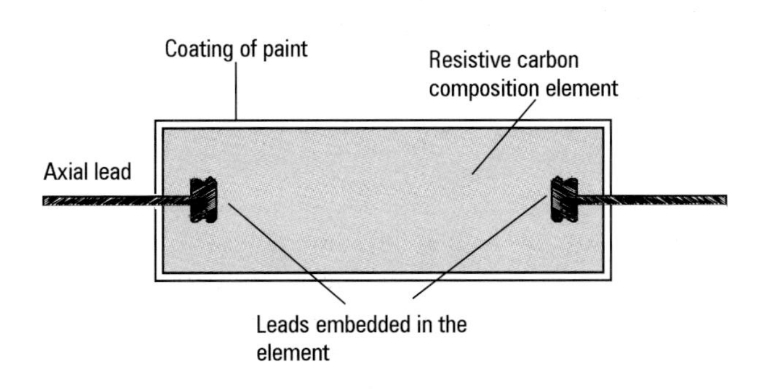


*Figure 5.31 Carbon composition resistor*

### Film resistors

Film resistors (Figure 5.32) are made by first placing a film of resistive material around a ceramic tube or rod. The resistive material in this case can be carbon, carbon-boron or some metallic oxide. A helical groove is cut into the film coating so that a resistive spiral track is left. Contact to the leads is usually by metal caps forced over the ends of the ceramic rod so that they are in contact with the film. The leads are then soldered or spot welded to the caps.

The component is then coated in a suitable lacquer for protection.

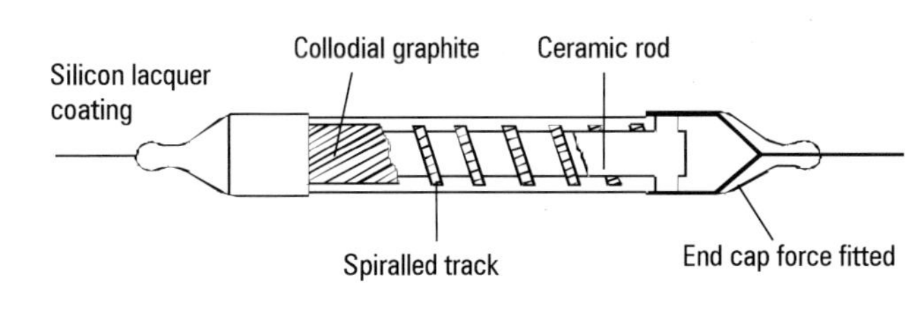


*Figure 5.32 Film resistor*

### Wire-wound resistors

As the name implies these are made by winding a length of resistance wire or tape on a ceramic or fibreglass bobbin. The ends of the wire are usually anchored to terminations on the end of the bobbin. The whole assembly is usually encapsulated in an impervious coat of vitreous enamel, ceramic or aluminium clad body (Figure 5.33). The resistance element usually consists of nickel–chromium alloy wire (nichrome), although in precision types this would be Eureka wire.

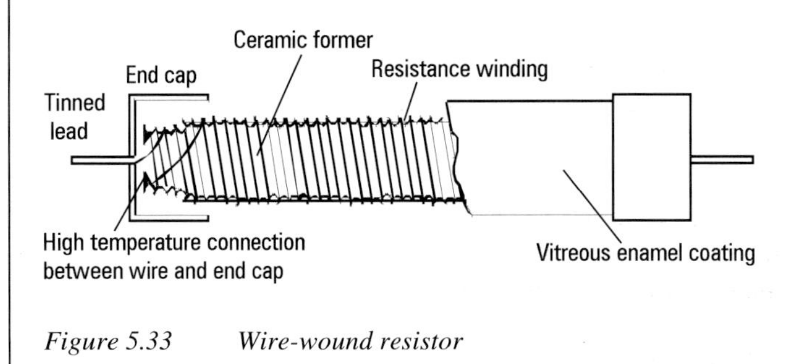


*Figure 5.33 Wire-wound resistor*

## Resistor identification

Resistors are normally easy to recognise from their shape and the presence of colour bands or number codes.

### Colour code

The colour code for fixed value resistors assigns a numerical value to each one of a range of colours as follows.

*Table 5.2*

| Colour | Value |
|---|---|
| black | 0 |
| brown | 1 |
| red | 2 |
| orange | 3 |
| yellow | 4 |
| green | 5 |
| blue | 6 |
| violet | 7 |
| grey | 8 |
| white | 9 |

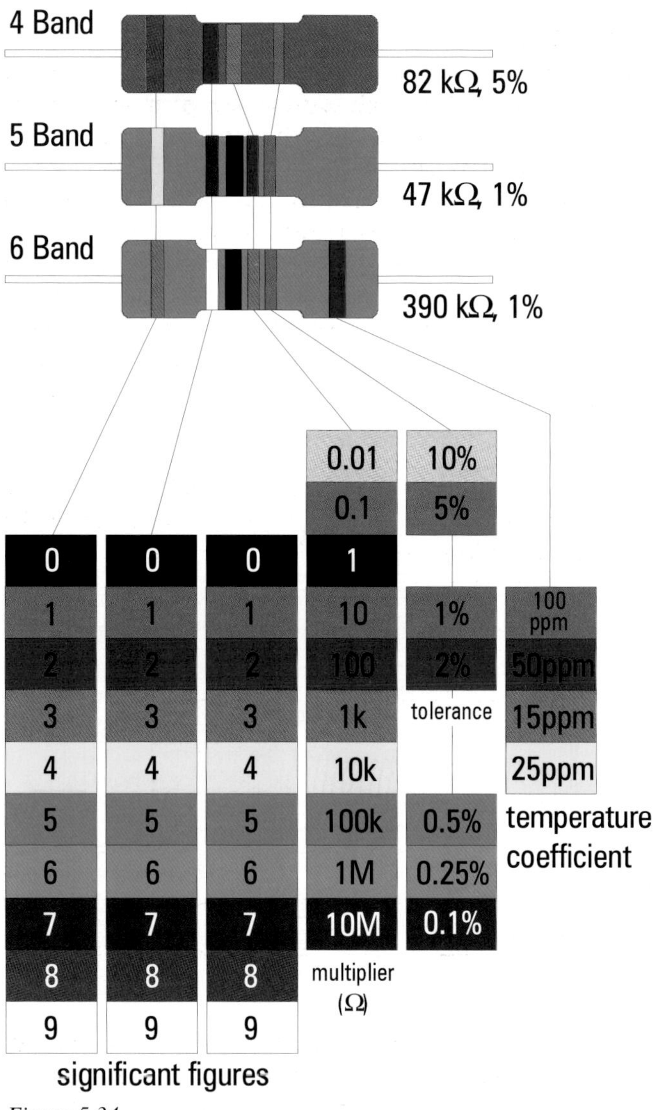


*Figure 5.34*

The method of applying the colour code involves reading off the colours one at a time and translating them into digits or multiplying factors. It will be found that the colour bands are grouped towards one end, and it is from this end that the colours must be read.

For values over 10 ohms the first two bands give a numerical value between 10 and 99 and the third gives the number of zeros to go with this number.

## *Example*

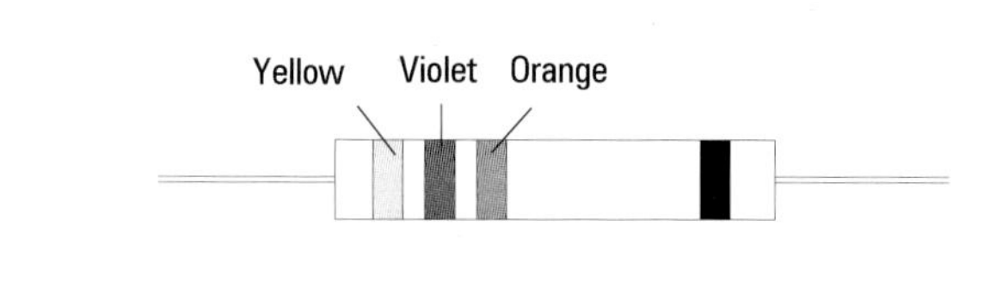


*Figure 5.35*

In Figure 5.35:

| | | | |
|---|---|---|---|
| **First Band** | Yellow | First digit | 4 |
| **Second Band** | Violet | Second digit | 7 |
| **Third Band** | Orange | No. of zeros | 3 |

i.e. 47 000 ohms

This is quite a simple technique, but must be practised regularly in order to evaluate resistors with speed and accuracy.

Several rhyming aids to memory have been devised to assist with the translation of the colour code, but frequent practice is by far the best method of learning to use the code.

Values under 10 ohms are recognisable by their third colour band being gold or silver. The value of the resistor is found in the normal way, using the first two colour bands, but the third band effectively moves the decimal point towards the left:

**Gold** one place
**Silver** two places

## *Examples using the colour code (Figures 5.36–5.38):*

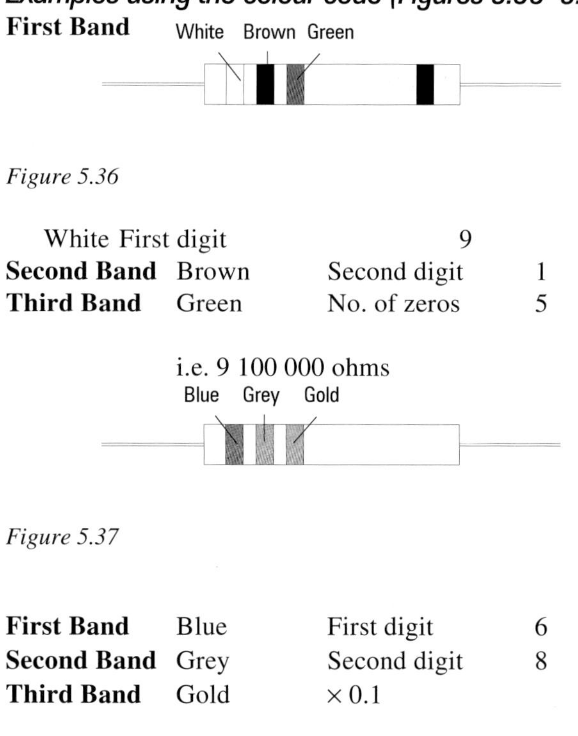

**First Band**



*Figure 5.36*

| | | | |
|---|---|---|---|
| White | First digit | | 9 |
| **Second Band** | Brown | Second digit | 1 |
| **Third Band** | Green | No. of zeros | 5 |

i.e. 9 100 000 ohms



*Figure 5.37*

| | | | |
|---|---|---|---|
| **First Band** | Blue | First digit | 6 |
| **Second Band** | Grey | Second digit | 8 |
| **Third Band** | Gold | $\times 0.1$ | |

i.e. $68 \times 0.1 = 6.8$ ohms

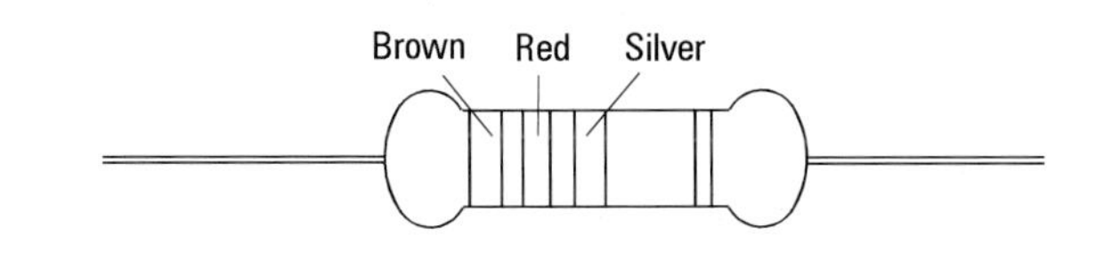


*Figure 5.38*

| | | | |
|---|---|---|---|
| **First Band** | Brown | First digit | 1 |
| **Second Band** | Red | Second digit | 2 |
| **Third Band** | Silver | $\times 0.01$ | |

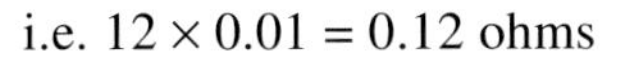

i.e. $12 \times 0.01 = 0.12$ ohms

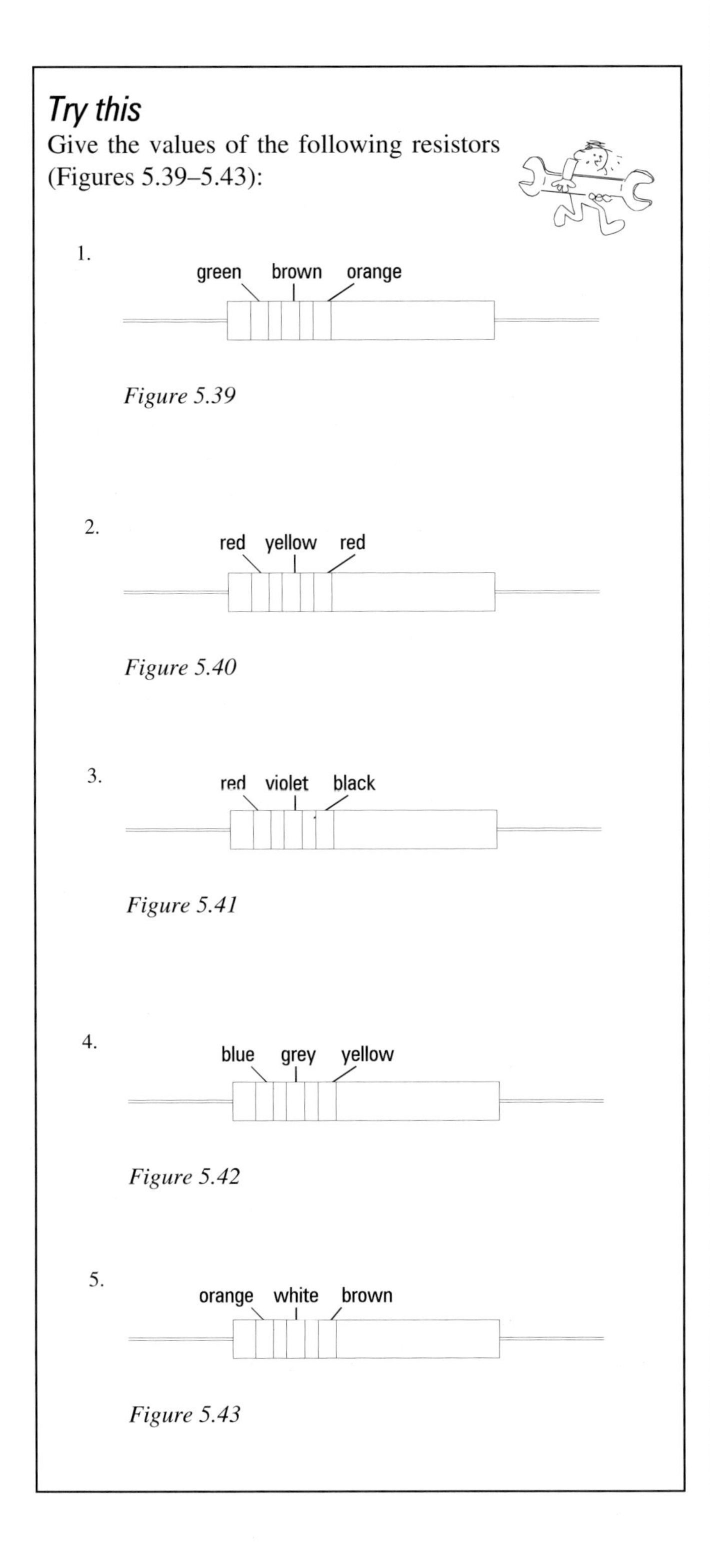


## Tolerance

Care must always be used when selecting a resistor for a particular use. Although all resistors are given a value, the accuracy of that stated value must be known. If an application demands very accurate values then resistors made to very fine tolerances must be used. However, if it is not so important to have very precise values then resistors with wider tolerances, say up to 20%, can be used. Resistors are manufactured in such a way that their value, instead of being exactly as stated, lies within an accepted range centred around that value. The extent of this range either side of the nominal value is indicated by the fourth band, which indicates the tolerance and is normally given as a percentage.

You will have noticed that in Figures 5.35, 5.36 and 5.38 there is a fourth band. This band indicates the tolerance of the resistor. The resistor in Figure 5.37 has no fourth band, but it still has a tolerance, as can be seen in Table 5.3

*Table 5.3*

| Colour | Tolerance ± |
|---|---|
| brown | 1% |
| red | 2% |
| gold | 5% |
| silver | 10% |
| no colour | 20% |

A 1 kΩ resistor for example may have a silver fourth band. This, as you see, indicates that it has a tolerance of ± 10%. This means that its actual value lies somewhere between 1000 + 10% (1100) and 1000 – 10% (900). A similar resistor having a gold fourth band could have a value anywhere between 950 ohms and 1050 ohms, its tolerance being ± 5%.

You will appreciate that resistors having smaller tolerances have actual values which are closer to the nominal value indicated by the first three colours.

Where a resistor has three colours only, this indicates that it has a tolerance of ± 20%.

## The system of preferred value

In many cases it is only necessary to specify a resistor which has a value that is within reasonable limits. Where greater accuracy is required a closer tolerance may be specified, but the same nominal value quoted. In other instances it may be necessary to obtain a resistor having a resistance which is not among those listed by the supplier.

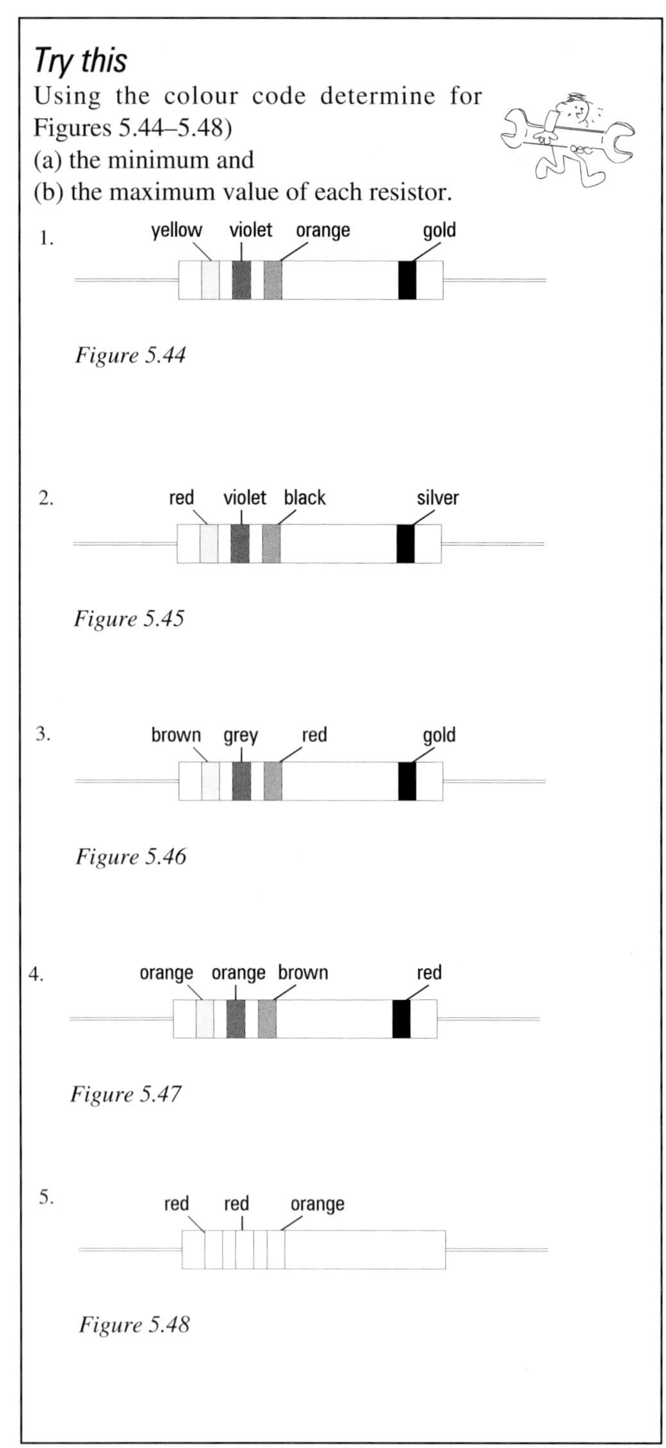


Figure 5.44

Figure 5.45

Figure 5.46

Figure 5.47

Figure 5.48

Due to the practice of stating tolerances, the actual value of any resistor will fall between the upper and lower limits of the tolerance range. Where large tolerances are given the width of the tolerance band contains a greater range of values. Effectively, where tolerances are large, fewer nominal values are required and as tolerances are reduced the number of nominal values required to cover the range is increased.

As the nominal resistance value increases the range between upper and lower values is increased, and in order to avoid unnecessary overlapping the interval between values becomes greater. This was how the preferred value system began. The E series numbers you will find in manufacturers' catalogues (as in E6) relate to the number of steps per range. The same preferred value system is repeated throughout all ranges.With improvements in manufacturing techniques this only applies up to a 5% overlap, but tighter tolerance resistors (2% or 1%) are now available.

*Table 5.4 Sample of preferred values*

| E6 20% tolerance | E12 10% tolerance | E24 5% tolerance | E96 1% tolerance |
|---|---|---|---|
| 10 | 10 | 10 | 100 |
| 15 | 12 | 11 | 102 |
| 22 | 15 | 12 | 105 |
| 33 | 18 | 13 | 107 |
| 47 | 22 | 14 | 110 |
| 68 | 27 | 15 | 113 |
| | 33 | 16 | 115 |
| | 39 | 17 | 118 |
| | 47 | 18 | 121 |
| | 56 | 20 | 124 |
| | 68 | 22 | 127 |
| | 82 | and so on | and so on |

High-precision resistors (for example 0.1%) also exist. Manufacturers and/or suppliers show the full range.

Although not common, some resistors will have 5 or even 6 bands. For example on a 6-band resistor the first 3 bands are significant digits, the fourth band is the multiplier, the fifth the tolerance and the sixth the temperature coefficient.

## Number Codes

Some resistors, notably wire-wound ceramic coated types or surface mount resistors, have their values printed on rather than being colour-coded and these are quite easy to evaluate since the resistance is clearly marked.

For example:

3.9k    Tol 5%

meaning that this is a 3900 ohm resistor with a tolerance of ± 5%.

This system has now been adapted because, in some instances, due to poor reproduction, the decimal point was omitted.

The BS 1852 resistance code is similar, but the decimal point is replaced by a letter:

R for ohms
K for kilohms
M for Megohms

## Example:

| | | |
|---|---|---|
| R56 | would mean | 0.56 Ω |
| 1R0 | would mean | 1 Ω |
| 27R | would mean | 27 Ω |
| 6K8 | would mean | 6.8 kΩ |
| 470K | would mean | 470 kΩ |
| 3M3 | would mean | 3.3 MΩ |

After this is added a second letter to indicate tolerance.

| | | |
|---|---|---|
| F | = | 1% |
| G | = | 2% |
| J | = | 5% |
| K | = | 10% |
| M | = | 20% |

thus

| | | |
|---|---|---|
| 8M2M | = 8.2 Megohm | ± 20% |
| 22 KJ | = 22 kilohms | ± 5% |
| 4R7K | = 4.7 ohm | ± 10% |

*Figure 5.49* *3K9 would mean 3.9kΩ*

### Try this

Using the number codes give values for resistance and tolerance for each of the following:

1. 10RJ
2. 1K5K
3. 1M0M
4. 220KK
5. 680RM

Give the codes for the following values:

6. 1.5MΩ ± 5 %
7. 0.47 Ω ± 2 %
8. 10 MΩ ± 1 %
9. 330 Ω ± 10 %
10. 1800 KΩ ± 2 %

## Points to remember

The construction of resistors fall into three groups, carbon, film and ___________________

If the third band on a resistor is gold, what does this indicate?

On a six-band resistor what does each band represent?

## Self-assessment multi-choice questions

**Circle the correct answers in the grid below.**

1. When the third colour band on a resistor is gold or silver it indicates that its value is less than
   (a) 100 Ω,
   (b) 50 Ω
   (c) 25 Ω
   (d) 10 Ω
2. If a resistor has six colour bands the sixth band denotes the
   (a) the temperature coefficient
   (b) the tolerance
   (c) how many zeros after the second digit
   (d) power
3. The value of the resistor in Figure 5.50 would be
   (a) 24 Ω
   (b) 240 Ω
   (c) 2400 Ω
   (d) 24 kΩ

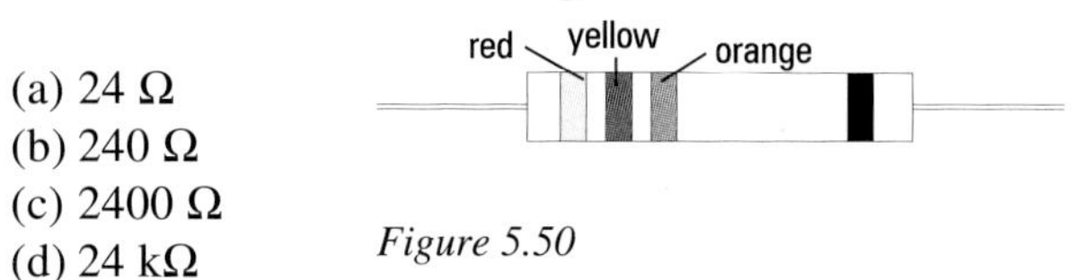


*Figure 5.50*

4. A resistor labelled 47RK would have a value of
   (a) 47 Ω ± 20%
   (b) 47 kΩ ± 5%
   (c) 47 Ω ± 10%
   (d) 47 kΩ ± 1%
5. A 56 kΩ resistor with a gold tolerance band could have a resistance between
   (a) 55 944 and 56 056 Ω
   (b) 53 200 and 58 800 Ω
   (c) 50 400 and 61 600 Ω
   (d) 44 800 and 67 200 Ω

### Answer grid

| | | | | |
|---|---|---|---|---|
| 1 | a | b | c | d |
| 2 | a | b | c | d |
| 3 | a | b | c | d |
| 4 | a | b | c | d |
| 5 | a | b | c | d |

# Part 3

## Capacitors

There are very few electronic circuits that do not contain a capacitor of some type or other. They can be fixed or variable and range in size from small plastic blobs the size of a solder drip to large cans that look as though they could hold food (Figures 5.51 and 5.52).

*Figure 5.51* *Sub-miniature ceramic capacitor*

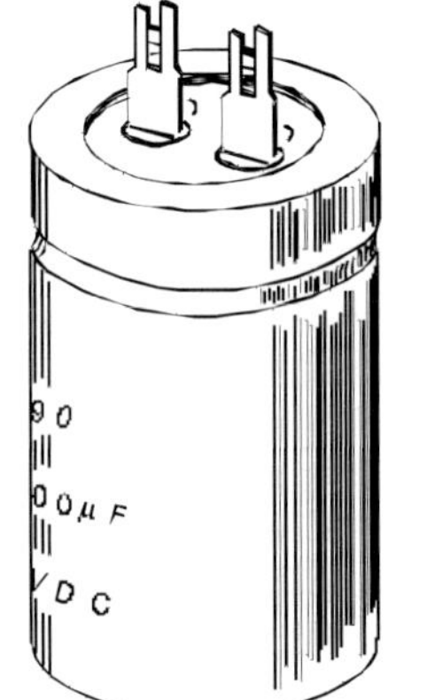

*Figure 5.52* *General-purpose electrolytic capacitor and symbol*

Capacitors have many different applications but we need first to look at what a capacitor is.

A capacitor consists basically of two metallic surfaces separated by an insulator known as a dielectric (Figure 5.53). As the plates are not in direct contact with each other they do not form a circuit in the same way that conductors and resistors do. They do, however, have an effect on current flow.

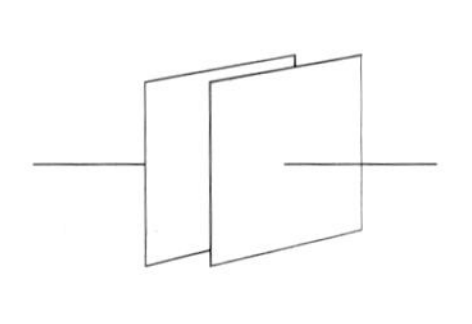

*Figure 5.53*

## Current flow

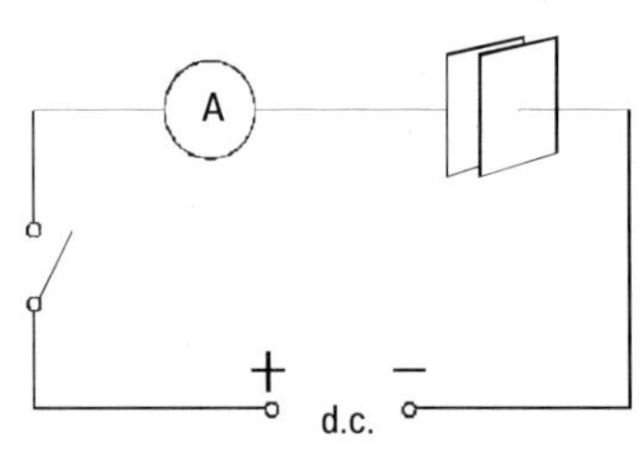


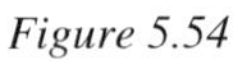

*Figure 5.54*

To examine this let's consider the basic circuit as shown in Figure 5.54, where a d.c. supply is connected to it. When the switch is closed there is a deflection on the meter which soon returns to zero.

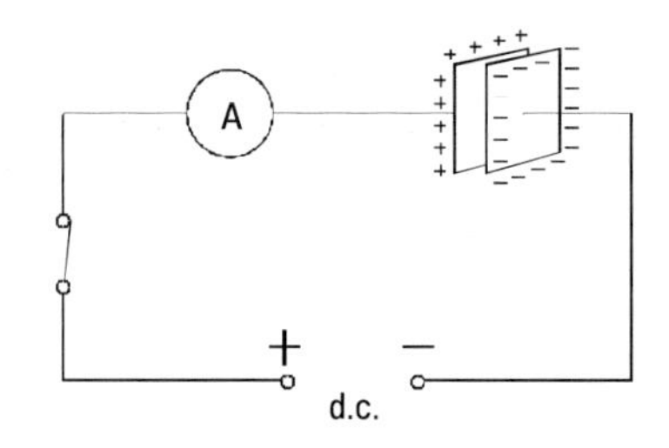


*Figure 5.55*

What has happened is that the positive terminal of the d.c. supply has built up a positive charge on the capacitor plate connected to it. In the same way a negative charge has built up on the negative plate (Figure 5.55).

## Voltage charge and discharge

These charges build up fast, but are not instantaneous: they follow a set pattern of "charge".

From the time the circuit is switched on a charge curve starts which is related to the charge voltage and time. The maximum voltage that the capacitor can take is that of the supply. This sets the voltage scale from zero volts to supply volts. The time scale is divided into sections called time constants. The pattern of charge is always the same.

In the first time period the charge goes from zero volts to approximately $^2/_3$ of the maximum voltage. In the next period the charge goes from the point that the first finished to about $^2/_3$ of what is left. This pattern continues until the capacitor is fully charged (Figure 5.56).

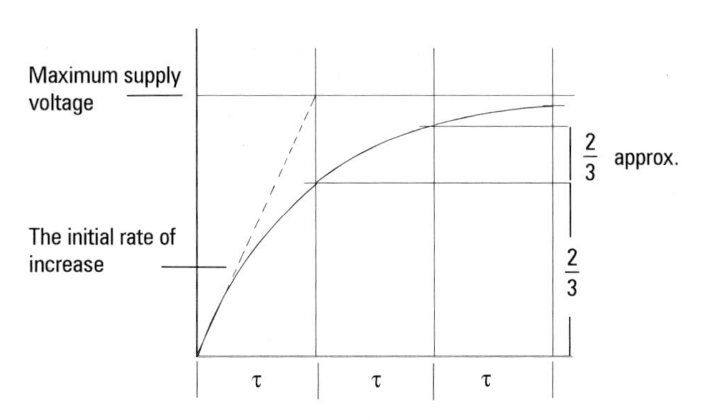


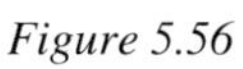

*Figure 5.56*

A time constant ($\tau$) is related to the value of the capacitance and the circuit resistance relevant to it (Figure 5.57).

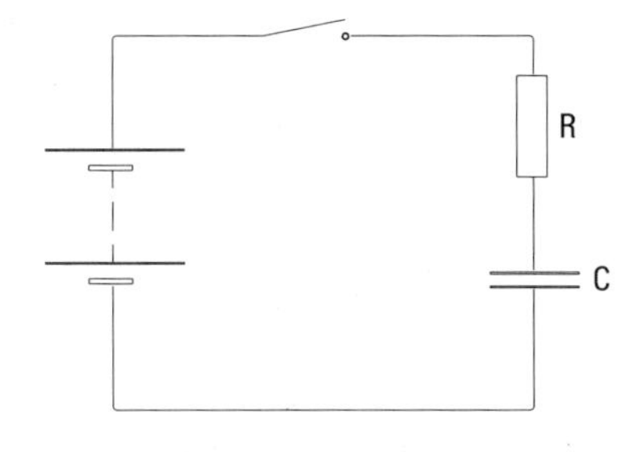


*Figure 5.57*

time constant ($\tau$) = $CR$ seconds

where $C$ is the capacitance and $R$ is the resistance.

When a capacitor is switched off and disconnected from any resistance, it will hold its voltage charge for some time. Care must always be taken when handling large capacitors, for they may still hold some charge.

On discharge a similar pattern of voltage occurs, but going from maximum to zero (Figure 5.58).

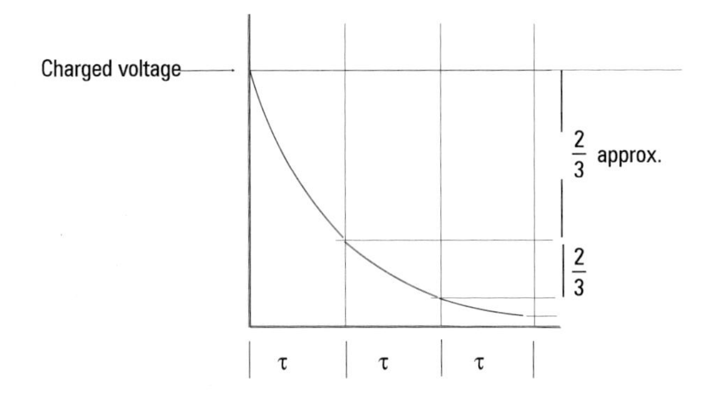


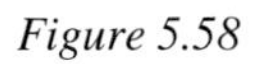
*Figure 5.58*

## The value of capacitance

The value of capacitance is measured in farads (F), but because of the very large size of this microfarads (μF) are usually used. Occasionally nanofarads and picofarads are used for very small values.

$$1\ \mu F = 10^{-6}\ F$$
$$1\ nF = 10^{-9}\ F$$
$$1\ pF = 10^{-12}\ F$$

These multiples must be used when carrying out calculations.

The charge stored in a capacitor is dependent on three main factors (Figure 5.59):

- the area of the facing plates (i): the larger the area the greater the capacitance
- the distance between the plates (ii): the less the distance the greater the capacitance
- the nature of the dielectric or spacing material

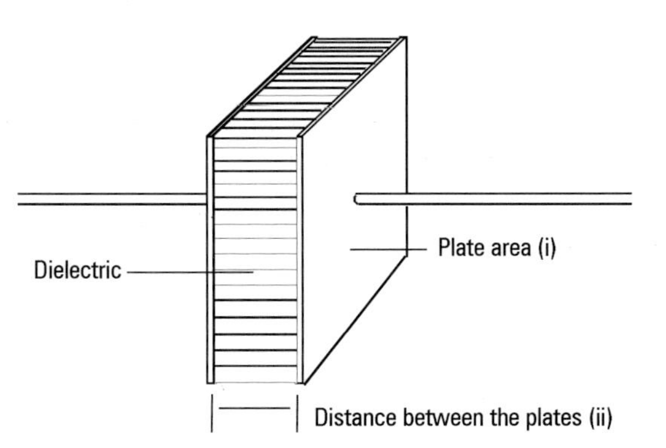


*Figure 5.59*

The charge stored by a capacitor is measured in coulombs ($Q$) and is related to the value of capacitance and the voltage applied to the capacitor.

$$\text{charge } (Q) = \text{voltage } (V) \times \text{farads } (C)$$

The voltage rating of a capacitor is very important, for if this is exceeded the dielectric may break down and the capacitor will short out.

The energy stored in a capacitor can be calculated from

$$\text{Energy } (W) = \frac{1}{2} C V^2$$

## Equivalent capacitance

Where it is not possible to replace a capacitor with one of the same value, several can be connected together to give an equivalent value.

We have seen that one way to increase capacitance is to increase the plate area. Connecting capacitors in parallel (Figure 5.60) has the same effect.

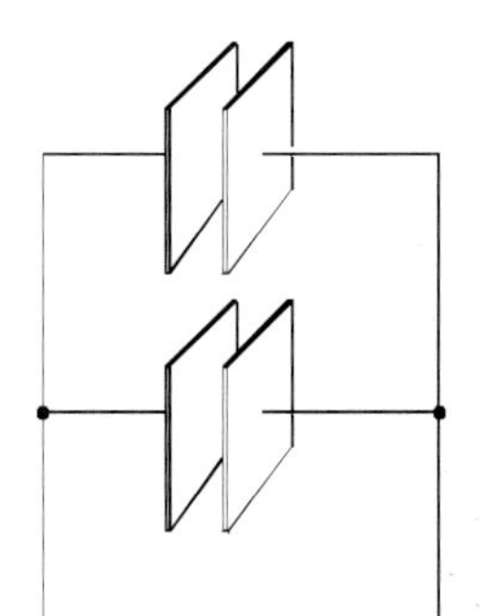

*Figure 5.60* *Capacitors connected in parallel*

By connecting two identical capacitors in parallel the plate area is effectively doubled and so is the capacitance.

When capacitors are connected in parallel the total capacitance increases.

$$C_{Total} = C_1 + C_2$$

If there are more than two capacitors in parallel, add them all together, as in

$$C_{Total} = C_1 + C_2 + C_3 + C_4 \ldots \text{ etc.}$$

To decrease the value of capacitance the space between the plates needs to be increased. When capacitors are connected in series (Figure 5.61) this is in effect what happens.

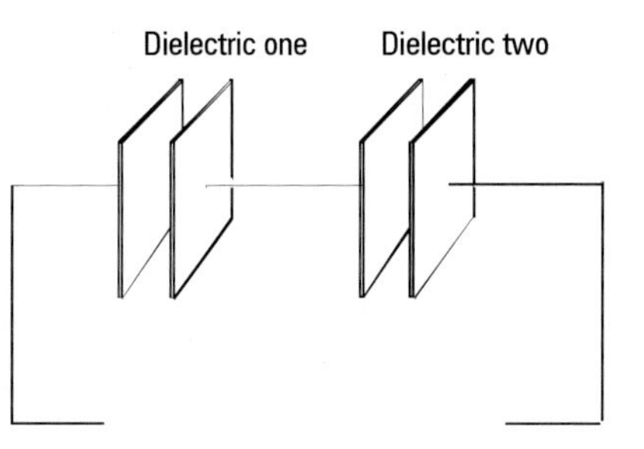


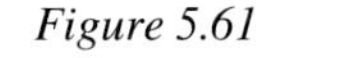
*Figure 5.61* *Capacitors connected in series*

By adding the dielectrics together the gap is doubled and the capacitance halved.

When capacitors are connected in series the total capacitance decreases.

$$\frac{1}{C_{Total}} = \frac{1}{C_1} + \frac{1}{C_2}$$

Again if there are more than 2 in series

$$\frac{1}{C_{Total}} = \frac{1}{C_1} + \frac{1}{C_2} + \frac{1}{C_3} + \frac{1}{C_4} \dots \text{etc.}$$

# Types of capacitor

To a large extent the application determines the type of capacitor that can be used. As with resistors, there are fixed and variable types of capacitor.

Fixed capacitors can be placed into three general classes related to their dielectrics.

| | |
|---|---|
| Low loss, high stability | – mica<br>– low K ceramic<br>– polystyrene |
| Medium loss, medium stability | – paper<br>– plastic film<br>– high K ceramic |
| Polarised capacitors | – electrolytic<br>– tantalum |

## Mica capacitors

Mica capacitors (Figure 5.62) have very low losses, even for frequencies up into the UHF (ultra-high frequency) ranges. They are suitable for use in RF (radio frequency) circuits up to 500 MHz and are recommended for use in oscillators and filters.

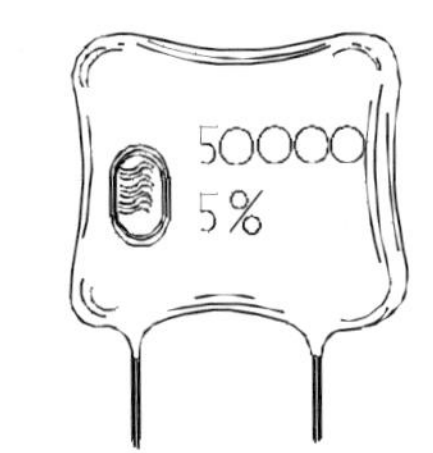


*Figure 5.62 Mica capacitor*

## Moulded mica (postage stamp)

The moulded mica capacitor (Figures 5.63 and 5.64) became commonly known as the "postage stamp" capacitor because of its shape and size.

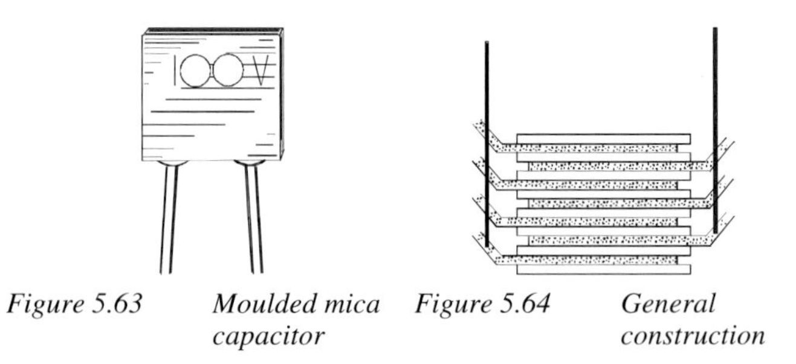

*Figure 5.63 Moulded mica capacitor* *Figure 5.64 General construction*

They are constructed of layers of foil interleaved with mica, often referred to as "stacked mica" or layers of metallized mica.

## Silver mica

Silver mica capacitors (Figure 5.65) are suitable for applications where high stability is required, as in tuned circuits and filters.

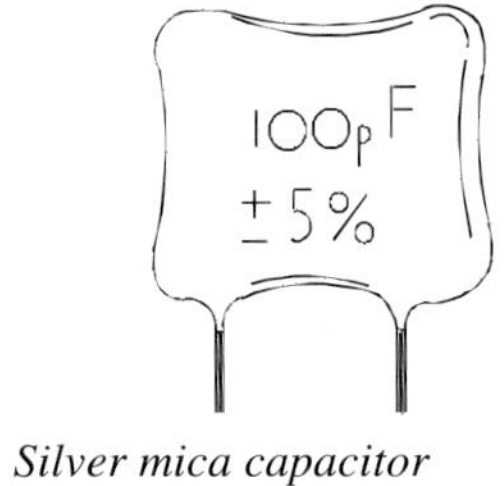


*Figure 5.65 Silver mica capacitor*

## Dipped mica

This is a type of encapsulation which is the result of dipping the capacitor stack into a resinous material below atmospheric pressure. Generally this improves the electrical characteristics and gives greater reliability over moulded types of encapsulation.

### *Try this*

Look up the voltage ranges of the listed types of capacitor in your electronics component catalogue.

| Type | Voltage range | | a.c./d.c. |
|---|---|---|---|
| | Maximum | Minimum | |
| | | | |
| | | | |
| | | | |
| | | | |
| | | | |
| | | | |
| | | | |

## Low-K ceramic capacitors

Low-K ceramics (Figure 5.66) have low loss and show small changes of capacitance with temperature. These are useful up to 1000 MHz and are made for both low-voltage and high-voltage applications.

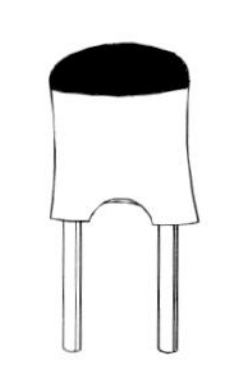

*Figure 5.66* *Low-K ceramic capacitor*

## Polystyrene capacitors

These are one of the plastic film type capacitors. They consist of alternate strips of foil and polystyrene film rolled up to form a tubular shape (Figure 5.67).

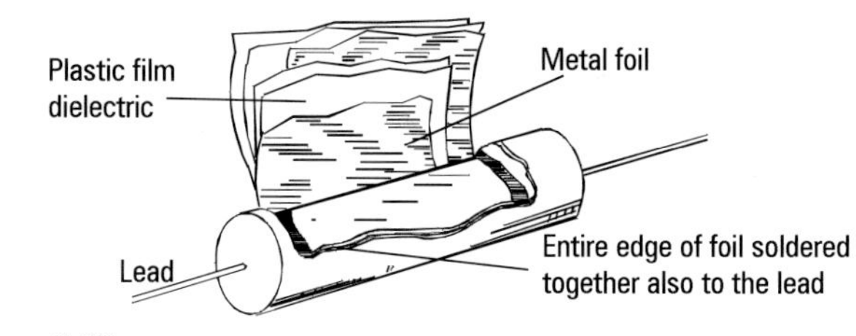


*Figure 5.67*

The alternate strips of foil are staggered and the entire edge of each foil is connected together and to the lead.

## Paper capacitor

This type of capacitor used to be very widely used, but the plastic film type capacitors have taken their place. The construction was similar to that of the plastic film type that has replaced them.

## Plastic film capacitors

These are widely used in the electronics industry due to their good reliability and relatively low cost. In addition to the polystyrene capacitor there are many types of plastic film in use. The plastics do, however, fall into three main categories:

- polystyrene
- polyester
- polycarbonate

The basic construction of plastic film capacitors is as described for the polystyrene capacitor. These are not always tubular in shape (Figure 5.68).

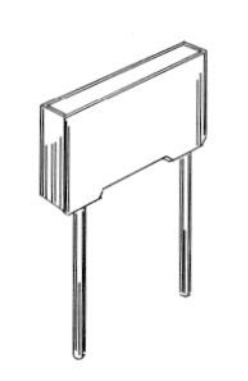

*Figure 5.68* *Plastic film capacitor*

They can be formed so as to be a flat rectangular shape so that more can be fitted into a given area of circuit.

## High-K ceramic capacitors

The advantage of this type of capacitor (Figure 5.69) is that they can supply large values of capacitance in a very small space. The capacitance of these tends to vary with temperature, and this limits their use to applications such as "d.c. blocking". They also change capacitance with applied d.c. and a.c. voltages, the capacitance decreasing with increasing voltage.

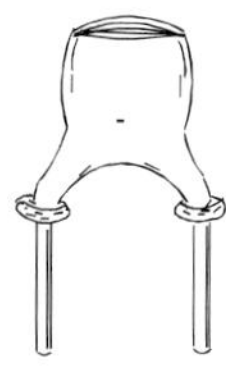

*Figure 5.69* *High-K ceramic capacitor*

## Electrolytic capacitors

These capacitors consist of two aluminium foils separated by an absorbent paper. Connections are fitted to the two foils and the whole assembly is rolled up and fitted tightly into an aluminium container which is hermetically sealed (Figure 5.70). The dielectric is formed electrolytically on the surface of one aluminium foil in the form of aluminium oxide. This foil acts as the positive plate, or anode, of the capacitor. The second plate, the electrolyte, is in contact with the other foil.

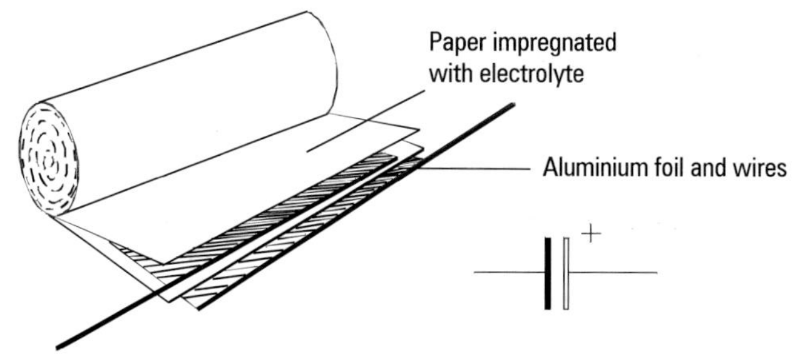


*Figure 5.70* *Electrolytic capacitor and symbol*

As the dielectric is formed by passing a direct current supply through the capacitor it means that the capacitor must always be connected to the same polarity. If it is connected around the wrong way or to an a.c. supply, the dielectric would break down and the capacitor would become a short circuit.

## Non-polarised electrolytics

These capacitors (Figure 5.71) are constructed by using several foils wound into one unit and connected "back-to-back". They are basically two electrolytic capacitors wound together in one container. This makes it possible to use electrolytic capacitors on a.c. supplies.

The construction of these means that they are much larger than ordinary electrolytic capacitors of the same value.

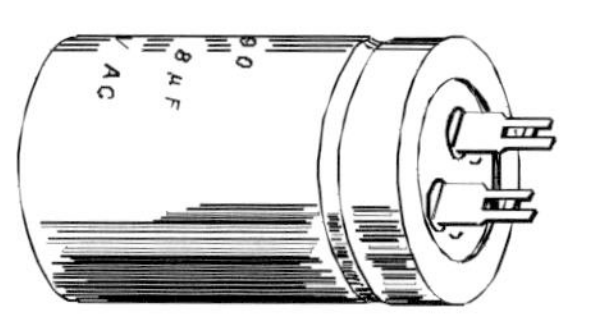

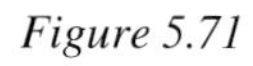

*Figure 5.71* *Non-polarised electrolytic capacitor*

*Table 5.5 Typical characteristics of some capacitors*

| | Type of capacitor | | | | | | |
|---|---|---|---|---|---|---|---|
| | Paper | Polyester | Polystyrene | Ceramic | | Electrolytic | |
| | Foil | Foil | | Disc | Monolithic | Foil | Tantalum |
| Temperature range (°C) | –30 to 100 | –40 to 100 | –40 to 70 | –55 to 125 | –55 to 125 | –20 to 80 | –40 to 150 |
| Capacitance range | 0.01 to 100 | 100 pF to 2.2 μF | 100 pF to 0.6 μF | 5 pF to 1 μF | 0.001 μF to 10 μF | 1 μF to 22 000 μF | 2.2 μF to 3500 μF |
| Voltage a.c. | 250 to 630 | 90 to 200 | | 63 to 250 | | | |
| Voltage d.c. | | 160 to 400 | 63 to 1000 | 63 to 10 000 | 63 to 450 | 6.3 to 500 | 200 to 1000 |

## Tantalum capacitors

The dielectric in these capacitors (Figure 5.72) is tantalum oxide. This is a much better dielectric than aluminium oxide and gives high values of capacitance in a relatively small space. The construction of these means they are polarised and must be connected in circuits to the correct polarity.

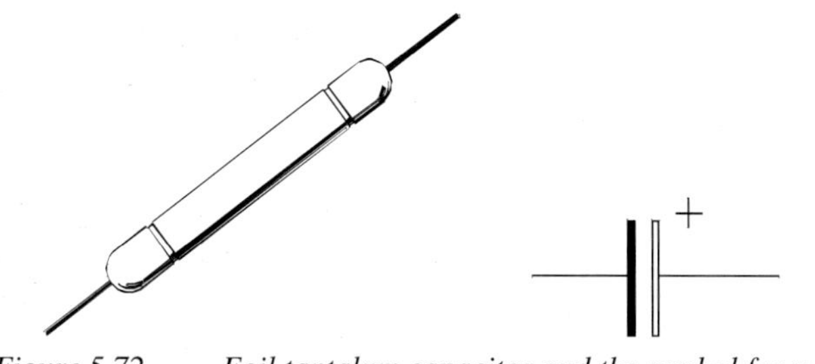

*Figure 5.72* *Foil tantalum capacitor and the symbol for a polarised electrolytic capacitor*

There are basically two different methods used to produce the two different types of tantalum capacitor.

These are

- the tantalum foil type
- the solid tantalum type

The tantalum foil type is constructed similarly to the electrolytic type but uses different materials.

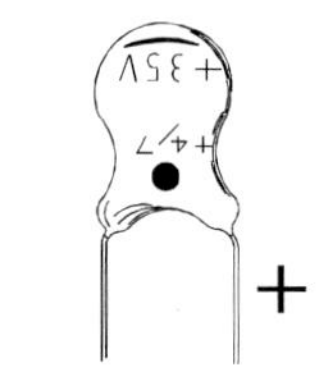

*Figure 5.73* *Solid tantalum capacitor*

The solid tantalum capacitors (Figure 5.73) consist of solid manganese dioxide as the electrolyte and a tantalum anode. The cathode is formed by the electrolyte being coated with graphite and silver. To ensure that the construction is not damaged it is usually encapsulated in epoxy resin.

## Variable capacitors

Variable capacitors (Figure 5.74) consist of one set of fixed plates and one set of moving plates. The greater the overlap of facing plates the greater the value of capacitance.

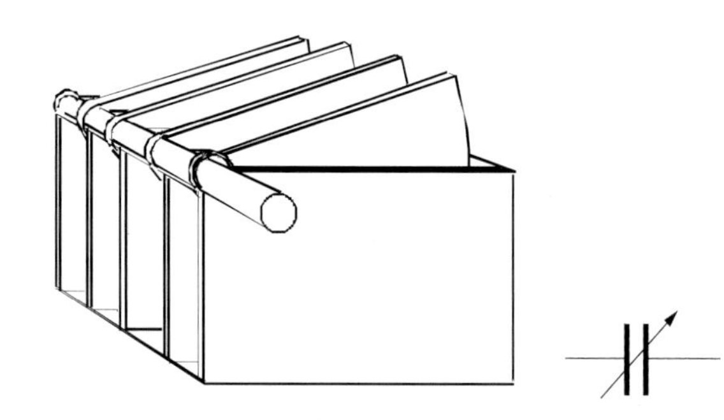

*Figure 5.74* *Variable capacitor and the symbol for a variable capacitor*

They can be made so that the user has access to adjust them, such as for radio tuning, or they can be trimmed and pre-set (Figure 5.75) by the manufacturer and sealed inside the equipment.

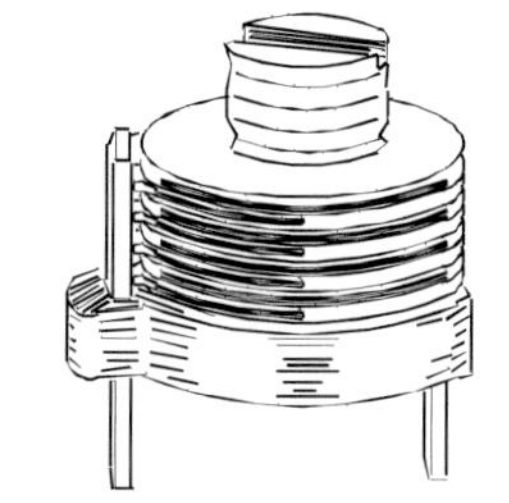

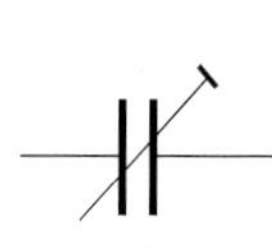

*Figure 5.75* *Trimmer and the symbol for a pre-set variable capacitor*

### *Remember*

Many of these types of capacitor are also available in surface mount form.

## Identification of capacitors

To identify a capacitor the following details must be known;

- the capacitance
- the working voltage
- the type of construction
- the polarity (if any)

The identification of capacitors is not easy because of the wide variation in shapes and sizes.

In the majority of instances the capacitance will be printed on the body of the capacitor. This often serves as a positive identification of the device as a capacitor.

As we have seen, the value is normally given in microfarads (μF), picofarads (pF) or, less commonly, in nanofarads (nF).

*Table 5.6 Capacitance conversion table*

| | | |
|---|---|---|
| 0.000 001 μF | = 0.001 nF | = 1 pF |
| 0.000 01 μF | = 0.01 nF | = 10 pF |
| 0.000 1 μF | = 0.1 nF | = 100 pF |
| 0.001 μF | = 1 nF | = 1000 pF |
| 0.01 μF | = 10 nF | = 10 000 pF |
| 0.1 μF | = 100 nF | = 100 000 pF |
| 1 μF | = 1000 nF | = 1 000 000 pF |
| 10 μF | = 10 000 nF | = 10 000 000 pF |
| 100 μF | = 100 000 nF | = 100 000 000 pF |

## Capacitor colour code

| | |
|---|---|
| **1st band** | 1st digit as for resistors |
| **2nd band** | 2nd digit as for resistors |
| | |
| **3rd band** | 3rd digit |
| Brown pF | × 10 |
| Red pF | × 100 |
| Orange pF | × 1000 |
| Yellow pF | × 10 000 |
| | |
| **4th band** | Tolerance |
| Black +/– | 20% |
| White +/– | 10% |
| | |
| **5th band** | Working voltage |
| Red | 250 V |
| Yellow | 400 V |
| Blue | 630 V |

For capacitors using the colour code the colours are read from top to bottom (Figure 5.76), the capacitance being given in pF.

This type of coding is found on general-purpose polycarbonate capacitors, giving them a slight "liquorice allsort" appearance.

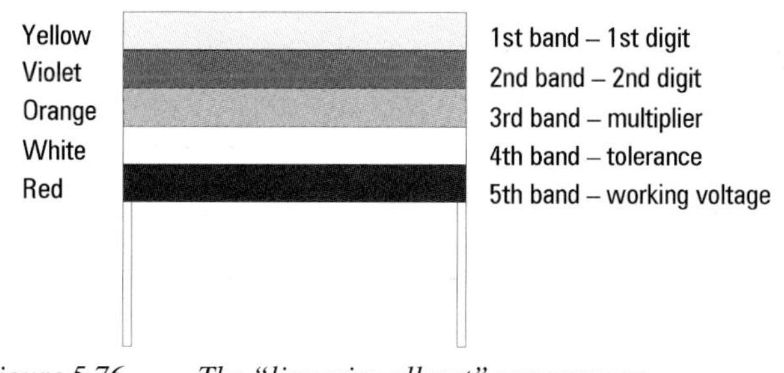


*Figure 5.76 The "liquorice allsort" appearance*

A similar colour coding system may be found on some types which have an appearance similar to resistors, but the coding uses dots rather than colour bands.

## Polarity

Once the size, type and voltage rating of a capacitor have been decided it only remains to ensure that its polarity is known.

As we have seen, some capacitors are constructed in such a way that if the device is operated with the wrong polarity its properties as a capacitor will be destroyed.

This is particularly true of electrolytic types.

The positive terminal must never be allowed to go negative, whether by wrong connection, supply reversal or by connection to an alternating voltage.

Polarity may be indicated by a + or – as appropriate. Electrolytics contained in metal cans may use the can as a negative connection. If no other marking is indicated but if it is still suspected that the capacitor is polarised, a slight indentation in the case will indicate the positive end (Figure 5.77).

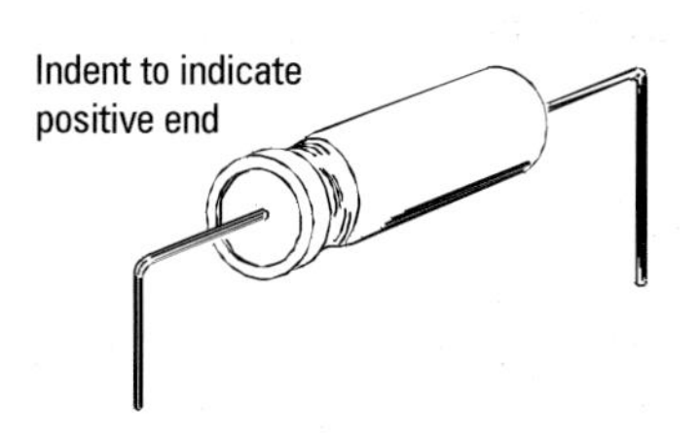


*Figure 5.77*

Tantalum capacitors having a spot on one side are polarised as shown in Figure 5.78. When the spot is facing you the right-hand lead is positive.

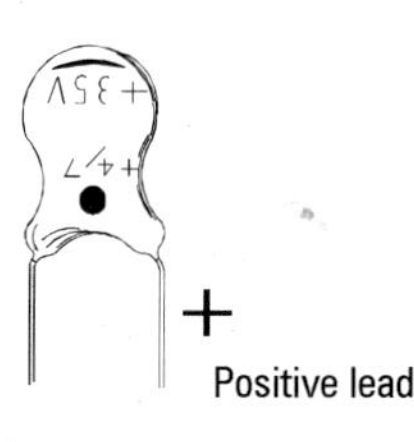


*Figure 5.78*

## *Try this*

Capacitors have to be connected to many different terminations. List below four different types of connection and suggest where they would be used.

1.

2.

3.

4.

## Suppression capacitors

A number of capacitors can be connected inside a single container to make up a suppression capacitor (Figures 5.79 and 5.80).

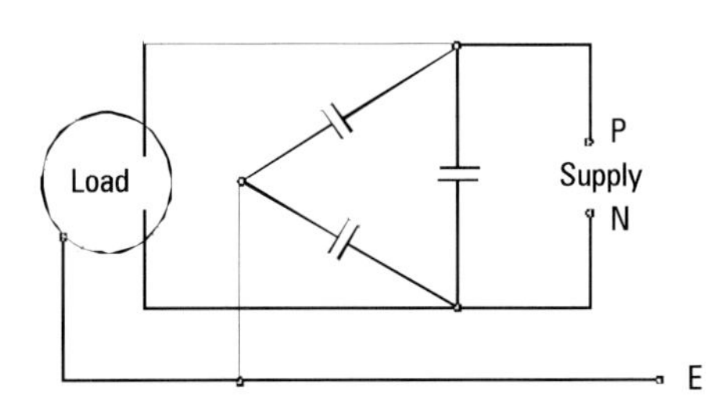


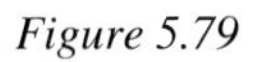
*Figure 5.79* *Suppression capacitor circuit*

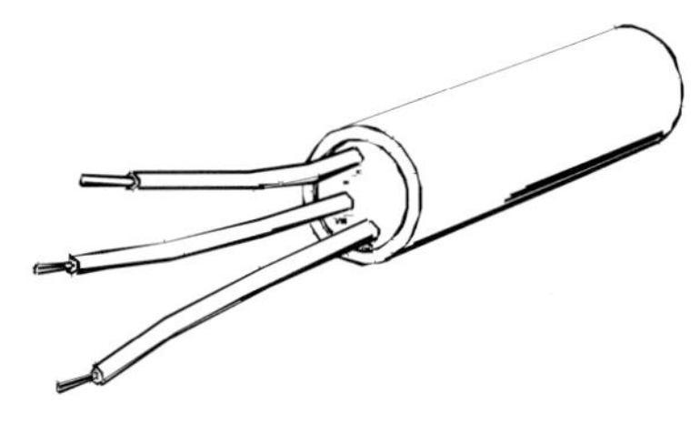

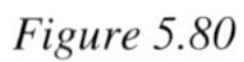
*Figure 5.80* *Suppression capacitor*

## The use of capacitors

As the rate of charge and discharge can be determined by the use of resistors, a capacitor/resistor arrangement can be used as a timing circuit. This usually forms part of a more complex circuit where one action depends on the result of another. We will be looking at a typical timing circuit later in this book.

On a.c. mains voltage circuits capacitors are often found being used for power factor correction to overcome the inductive effects of chokes or motor windings. In circuits where inductive loads are being switched, capacitors are connected across the contacts to "absorb" the spark and reduce radio interference (Figure 5.81). An example of this can be found in the starter switch of a fluorescent luminaire.

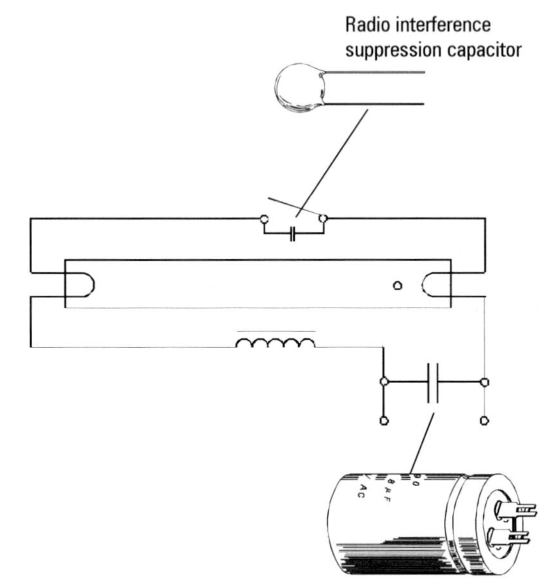


*Figure 5.81*

**Care must always be taken when handling large capacitors, for they may still hold a charge.**

In d.c. circuits capacitors are used slightly differently.

Often there is a ripple in the supply which is undesirable (Figure 5.82). A capacitor of the correct value and type can be used to reduce this to an acceptable level. In the smoothing circuit in Figure 5.83 the unwanted rise and fall in the output voltage is filtered out.

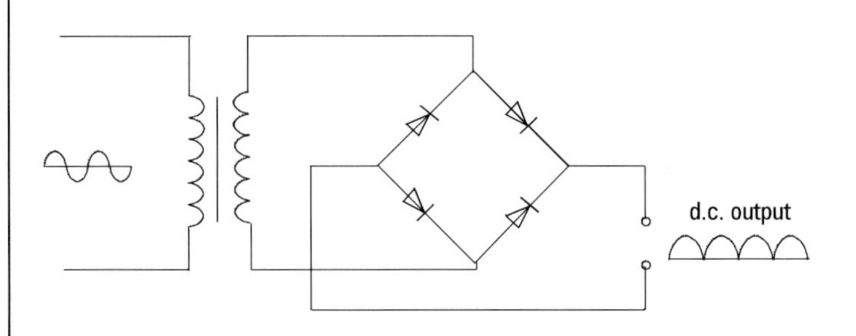


*Figure 5.82* *Full-wave rectified circuit without smoothing.*

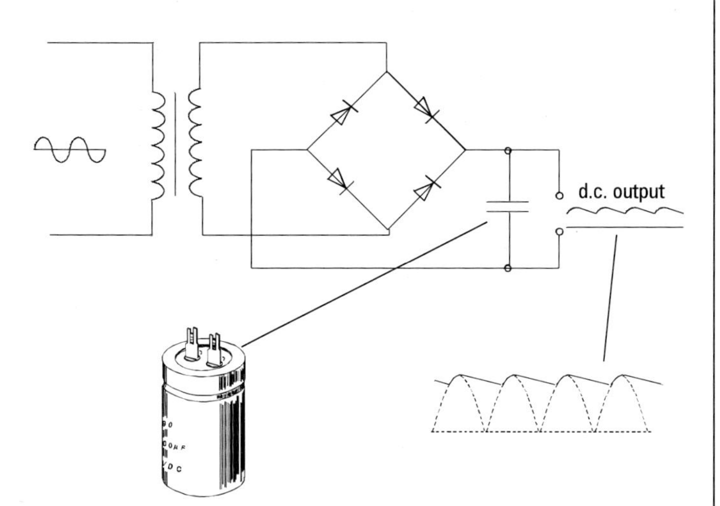


*Figure 5.83* *Full-wave rectified circuit with capacitor connected for smoothing*

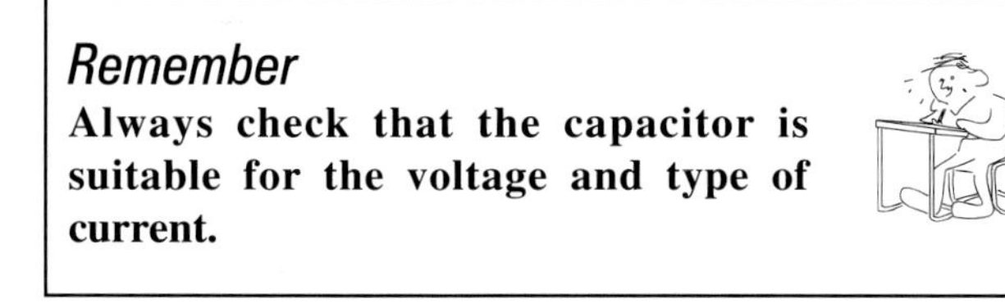

## Remember

**Always check that the capacitor is suitable for the voltage and type of current.**

## Points to remember ◀

Most electronic circuits will contain a capacitor of some type or other. The appearances of different types vary considerably and their physical size is not necessarily related to their value of capacitance.

Different types of capacitor have different applications and many cannot be interchanged with another of a different type. The voltage rating is very important, for if it is exceeded the capacitor may be ____________

Always check that capacitors are ________________ before handling them.

A slight indentation at one end of a capacitor will indicate that it is the ___________ end.

What type of capacitor is recommended for use in RF circuits?

What type of capacitor is recommended for use in tuned circuits that require high stability?

## Self-assessment multi-choice questions

**Circle the correct answers in the grid below**

1. In a capacitor two factors that can increase the capacitance are
   (a) smaller plates, thicker dielectric
   (b) smaller plates, thinner dielectric
   (c) larger plates, thicker dielectric
   (d) larger plates, thinner dielectric
2. (i) Two identical capacitors connected in parallel double the total capacitance.
   (ii) The energy stored in a capacitor can be calculated from $W = \frac{1}{2}CV^2$.
   (a) only statement (i) is correct
   (b) only statement (ii) is correct
   (c) both statements are correct
   (d) neither statement is correct
3. A polarised electrolytic capacitor can be used on
   (a) a.c. only
   (b) d.c. only
   (c) both a.c. and d.c.
   (d) both a.c. and full-wave rectified current
4. The capacitor connected across the contacts of a starter switch in a fluorescent luminaire circuit is to
   (a) correct the power factor to the circuit
   (b) reduce the current taken by the lamp
   (c) counter the inductive effects of the choke
   (d) provide radio interference suppression
5. 1 nF is equal to
   (a) 0.001 μF
   (b) 100 μF
   (c) 10 000 pF
   (d) 0.01 pF

### Answer grid

| | | | | |
|---|---|---|---|---|
| 1 | a | b | c | d |
| 2 | a | b | c | d |
| 3 | a | b | c | d |
| 4 | a | b | c | d |
| 5 | a | b | c | d |

# Part 4

## Inductors and transformers

There are two types of inductance: self-inductance and mutual inductance.

Self-inductance relates to single coils and inductors or chokes. Mutual inductance applies where there are two or more coils, such as in a double-wound transformer. The two coils are electrically isolated but magnetically coupled, such that a change in magnetic flux driven by one coil induces a current in the other. An inductor is shown in Figure 5.84.

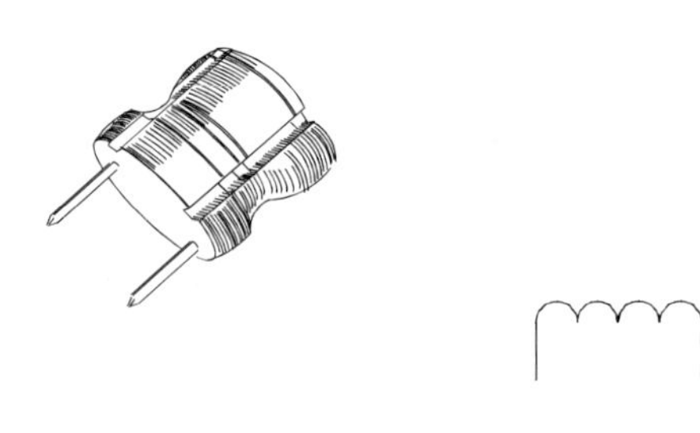

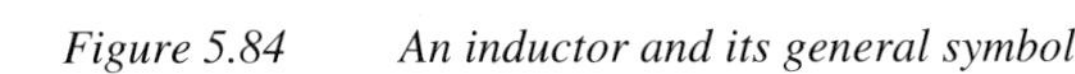

*Figure 5.84 An inductor and its general symbol*

### Low-frequency inductors

A low frequency inductor basically consists of a coil wound on a bobbin and an iron core built up of silicon steel laminations (Figures 5.85–5.87). The iron core is used to create a magnetic path throughout the coil.

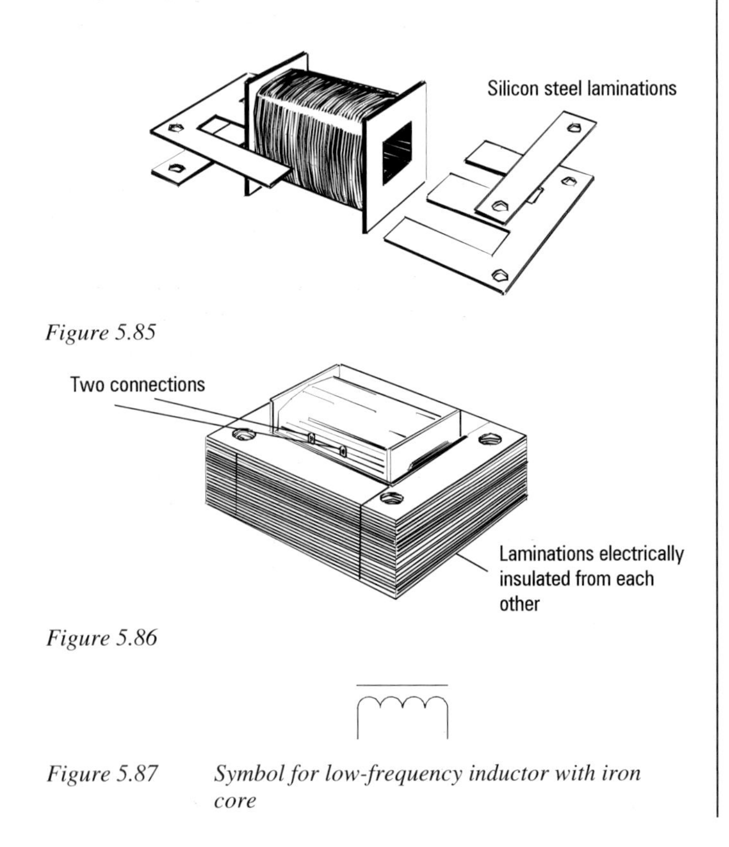


*Figure 5.85*

*Figure 5.86*

*Figure 5.87 Symbol for low-frequency inductor with iron core*

### High-frequency inductors

A high-frequency inductor is a coil wound over a core which is made up of a powder which is a good magnetic conductor. Each grain of powder is coated with an electrical insulator and then it is pressed and moulded into the required shape (Figure 5.88).

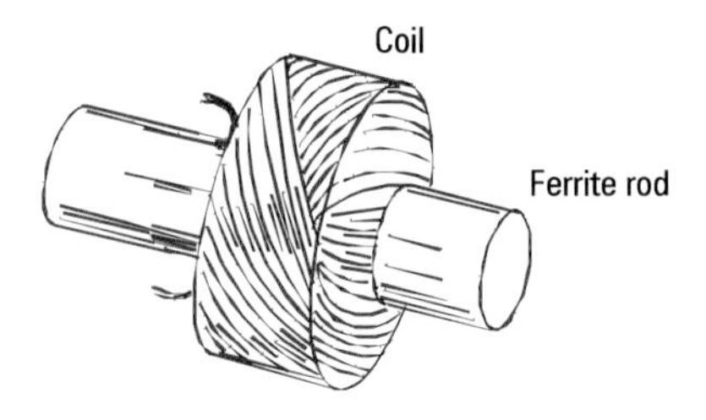


*Figure 5.88 High-frequency inductor*

The powder may be an iron dust or ferrite material. Ferrite is made up of ferric oxide combined with other oxides, such as nickel. Where the inductance of a coil is to remain set the coil or coils are fixed to a ferrite rod. If, however, the inductance of the coil has to be adjusted, a threaded iron dust core is used (Figure 5.89).

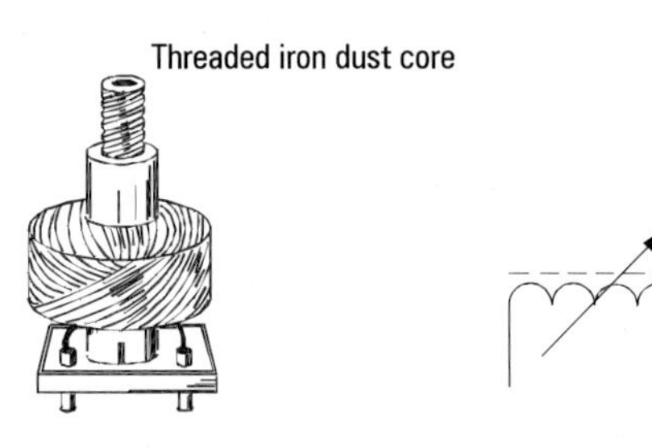


*Figure 5.89 Adjustable high-frequency inductor with ferrite core and its symbol*

Some inductors are air-cored. These inductors have small inductances so they are suitable for high-frequency circuits.

### Action

As an inductor is a coil of wire it has resistance, and when connected to a d.c. supply it will act as a resistor. However, when a coil is connected to an a.c. supply it becomes an inductor and takes on additional properties as it reacts with the alternating current. Inductance is measured in henrys, the quantity symbol is $L$ and the unit symbol is H.

The alternating current creates the effect of a continuously changing magnetic field inside the coil. This effect reacts with the flow of current and opposes it. This means that when the coil is connected to a d.c. supply it is only the resistance of the coil that limits the flow of current, but when connected to an a.c. supply there is also reactance due to the inductance of the coil. Inductive reactance, as it is called, is measured in ohms; the quantity symbol is $X_L$ and the unit symbol Ω.

To show the effect of inductance a graph can be plotted similar to that for capacitance, but in this case current, rather than voltage, is plotted against time (Figure 5.90).

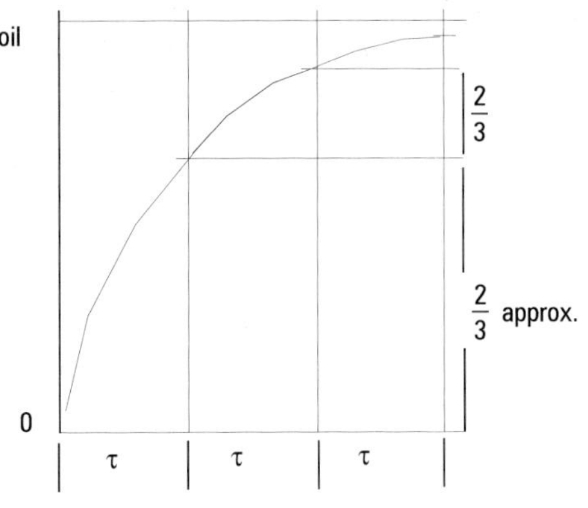


*Figure 5.90*

One time constant is calculated from

$$\tau = \frac{L}{R} \text{ seconds}$$

where $L$ is the inductance in henrys and $R$ is the resistance in ohms.

The fact that it takes time to build up to the maximum current is very important, for when the coil is connected to an a.c. supply the current/time can be calculated and adjusted so that a smoothing effect can be produced in the alternating current.

If the coil is suddenly switched off the magnetic field collapses and a high voltage is induced across the circuit.

The inductance of a coil can be increased by

- increasing the number of turns of wire on the coil
- increasing the iron core in the coil. How this is achieved depends on the application of the inductor.

Where a number of inductors have to be connected together to form an equivalent inductance they follow the same rules as for resistors.

To **increase** inductance they are connected in **series**.

To **decrease** inductance but increase the current rating they are connected in **parallel**.

*Try this*

**List five applications for chokes or inductors.**

*Remember*

**An inductor is a coil with a magnetic conductor through the centre.**
**The unit name for induction is the henry, the unit symbol is H and the quantity symbol *L*.**

## Transformers

The transformer (Figures 5.91 and 5.92) in its basic form is two independent coils of wire connected only by a magnetic conductor. These rely on mutual inductance from one coil to another. One coil is connected to an a.c. source, which may be a signal from an aerial or amplifier, or a mains supply. This a.c. is transferred and transformed into the second coil.

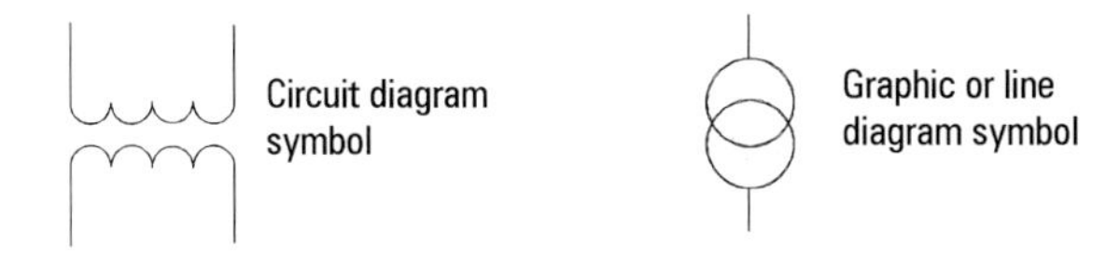


*Figure 5.91 General symbols for transformers*

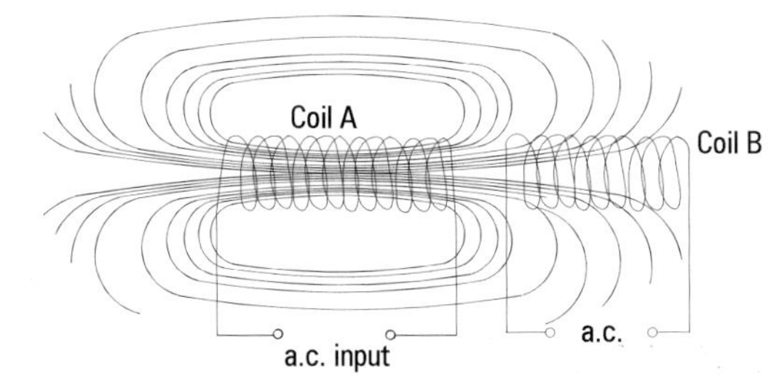


*Figure 5.92*

As the current in coil A builds up it produces a magnetic field which overlaps into coil B. Being a.c., the current is continuously changing direction so the magnetic field is continuously moving through coil B. This induces an e.m.f. into coil B which becomes the a.c. output.

## Low-frequency transformers

To improve the link between the two coils a magnetic conductor is used, such as silicon steel (Figures 5.93 and 5.94).

As we have seen previously, if the magnetic circuit was made of a solid piece of metal it would get hot due to the continuously changing magnetic field, so the core is built up of laminations pressed out of sheet silicon steel material.

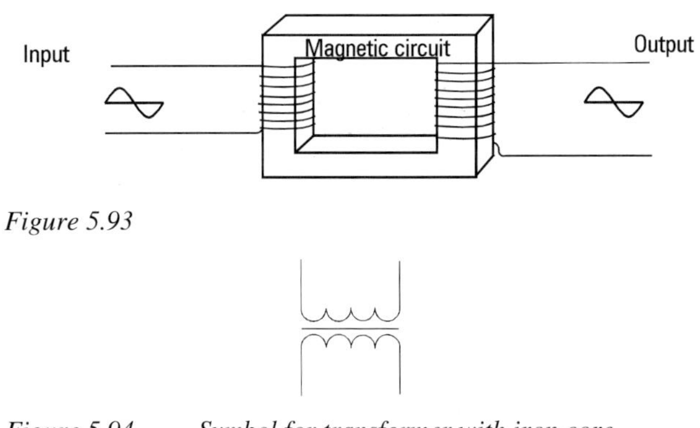


*Figure 5.93*

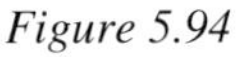

*Figure 5.94 Symbol for transformer with iron core*

## Turns ratio

The output of a transformer can be determined by the turns ratio of the two coils.

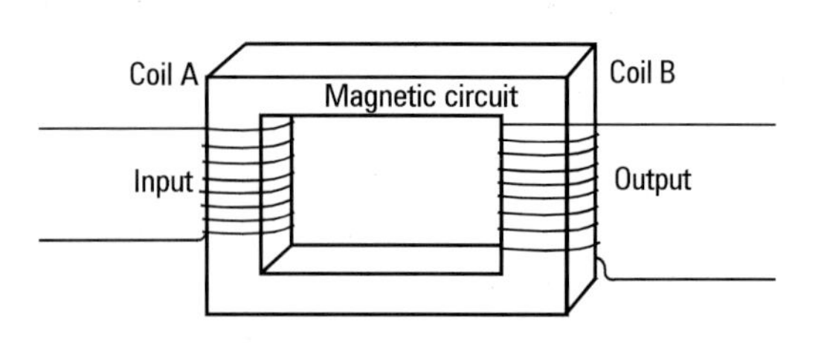


*Figure 5.95*

For example (Figure 5.95), if coil A had 4000 turns and coil B only 2000 the transformer would have a ratio of 4000/2000 or 2:1. This means that if 200 volts a.c. is connected to coil A the output of coil B will be in the ratio of 2:1, in this case 100 volts.

## RF transformers

Transformers used on radio frequency (RF) usually have adjustable iron dust cores (Figure 5.96) instead of the laminated steel type. This is because they often form part of a tuning circuit and need to be adjusted to react with other components. These are usually enclosed in an aluminium can so as to screen them from other parts of the circuit (Figure 5.97). The symbol for an RF transformer is shown in Figure 5.98.

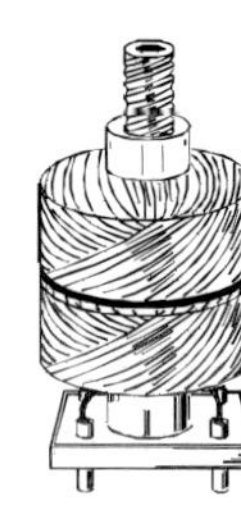

*Figure 5.96* *RF transformer with adjustable iron core*

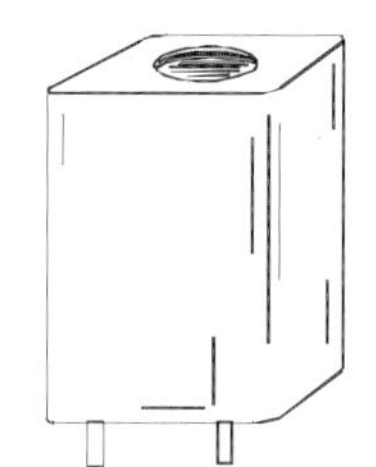

*Figure 5.97* *Aluminium can to screen RF transformer*

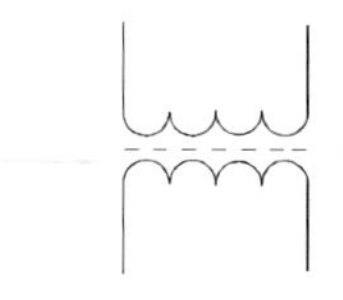

*Figure 5.98* *Symbol for RF transformer with ferrite core*

## Identification of inductors and transformers

In appearance there is little difference between a transformer and a low-frequency inductor, but further investigation with a continuity tester will reveal whether the device has two or more separate windings, which will identify it as a transformer (Figure 5.99).

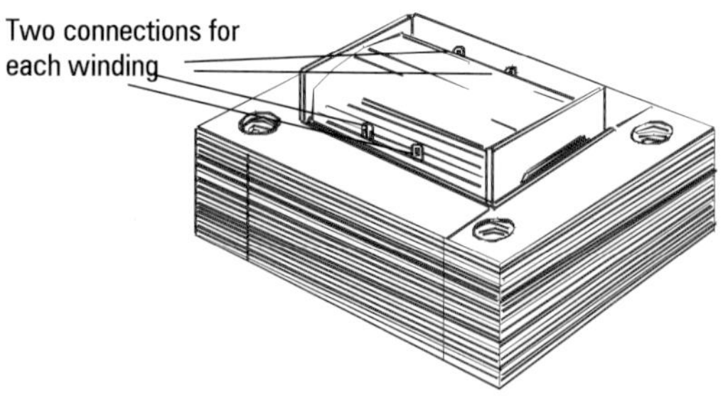


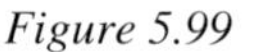

*Figure 5.99*

A choke or inductor should have its inductance in henrys (H) or millihenrys (mH) clearly marked. This indication should rule out the possibility of the device being an autotransformer. Autotransformers (Figure 5.100) have a single tapped winding as compared with the two separate windings of a double-wound transformer.

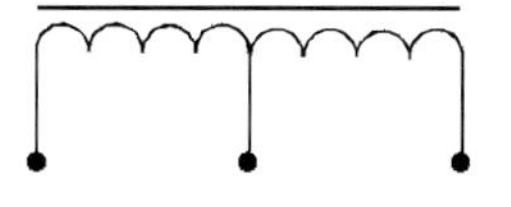

*Figure 5.100* *General symbol for an autotransformer*

## RF inductors

High-frequency inductors may be wound on formers and the core adjusted *in situ* to give the desired result (Figure 5.101).

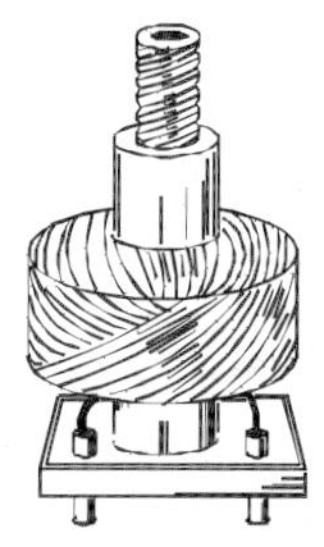

*Figure 5.101*

Fixed-value high-frequency inductors encapsulated in polypropylene are available in values up to 1000 μH.

*Remember*

**As a general guide chokes or inductors have two terminals, double wound transformers have four.**

## Relays

A relay is a device which allows one circuit to operate another. The simplest version of a relay consists of a set of contacts which either open or close, operated by a magnetic coil. More complex versions are available and are usually designated by the following:

| | |
|---|---|
| SP | single pole |
| DP | double pole |
| NO | normally open |
| NC | normally closed |
| ST | single throw (on–off) (Form A) |
| DT | double throw (changeover) (Form C) |

or any combination of the above.

For example, SPNO is single pole, normally open.

The coils may be available in either d.c. or a.c. versions and their physical form varies with current switching ratings. The symbol for a relay is shown in Figure 5.103.

*Figure 5.102*

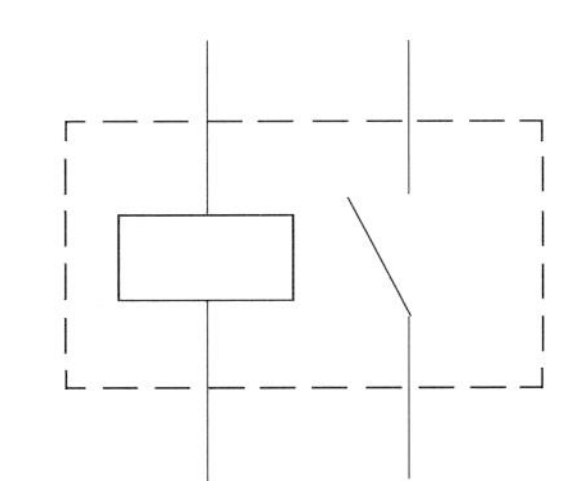

*Figure 5.103* *A typical relay symbol. Varieties of the symbols as regards the coils and contact forms can be found in BS3939. Manufacturers' catalogues may show alternative versions.*

### Points to remember

An inductor is basically a coil of wire which, when connected to an ____ supply has a continually moving magnetic field within it. The moving field induces an ________ within the field. This is known as self inductance.
A transformer is made up of two or more coils, one of which is connected to an a.c. supply. The coils have no electrical connection between them but are on a common magnetic circuit. The moving magnetic field resulting from the supply current induces an e.m.f. into the second coil. This is known as mutual inductance. The core of the magnetic circuit is made up of ________________________________________

### Self-assessment multi-choice questions

**Circle the correct answers in the grid below**

1. The magnetic core of a low-frequency transformer is
   (a) solid iron
   (b) silicon steel laminations
   (c) a ferrite rod
   (d) an iron dust rod
2. A low-frequency transformer supplied with 240 V has an output voltage of 9 V. If the primary winding has 4800 turns, the number of turns on the secondary winding would be
   (a) 2160
   (b) 180
   (c) 26.7
   (d) 20
3. The magnetic core of a high-frequency inductor could be made of
   (a) solid iron
   (b) silicon steel laminations
   (c) a ferrite rod
   (d) a steel rod
4. The aluminium can on an RF transformer is to
   (a) increase the magnetic circuit
   (b) increase the inductance of the coils
   (c) screen the coils from other components
   (d) make a neat presentation
5. (i) The number of terminals on a choke is usually only two.
   (ii) The number of terminals on a double-wound transformer is usually three.
   (a) only statement (i) is correct
   (b) only statement (ii) is correct
   (c) both statements are correct
   (d) neither statement is correct

### Answer grid

| | | | | |
|---|---|---|---|---|
| 1 | a | b | c | d |
| 2 | a | b | c | d |
| 3 | a | b | c | d |
| 4 | a | b | c | d |
| 5 | a | b | c | d |

# Part 5

## Semiconductor material

A semiconductor material is one whose resistivity lies between that of a perfect insulator and that of a good conductor.

These materials can be manufactured into devices such as diodes, transistors, thyristors and integrated circuits.

*Figure 5.104*

### Construction and action

A semiconductor diode is basically a piece of n-type and a piece of p-type semiconductor joined together.

N-type material is made by treating a semiconductor crystal such as germanium or silicon (Figure 5.105) with an impurity such as arsenic or antimony so that free electrons are introduced into the pure crystal (Figure 5.106).

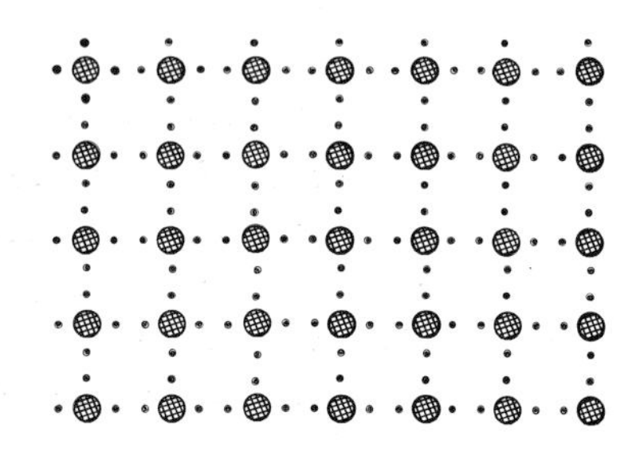

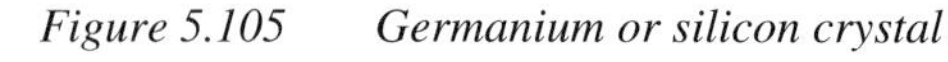

*Figure 5.105 Germanium or silicon crystal*

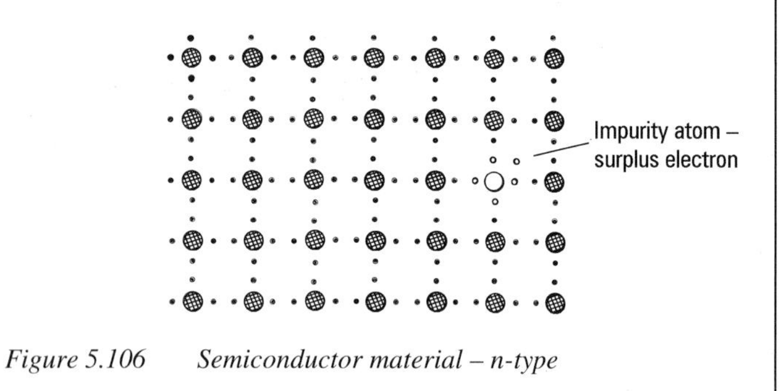


*Figure 5.106 Semiconductor material – n-type*

Although the overall charge of the crystal remains zero as each individual atom is electrically neutral, there is a free electron in the structure. As electrons are negatively charged this material becomes n-type material.

P-type material is also made by treating the semiconductor with an impurity. In this case the pure crystal is "doped" with an atom which has fewer electrons than those in the crystal. This creates a situation where there is effectively a missing electron, usually called a hole (Figure 5.107).

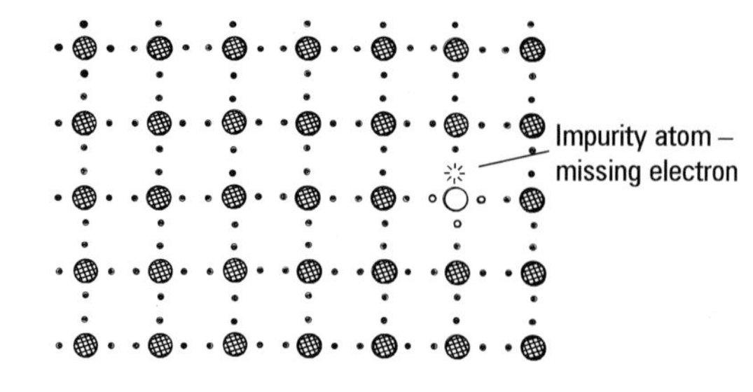


*Figure 5.107 Semiconductor material – p-type*

When the junction is formed between the n-type and p-type materials (Figure 5.108) there is an electron drift across the junction. Electrons from the n-type material move into the "holes" in the p-type, each leaving a positively charged nucleus. The result is that near the junction, the n-type material is positively charged and the p-type material is negatively charged (Figure 5.109).

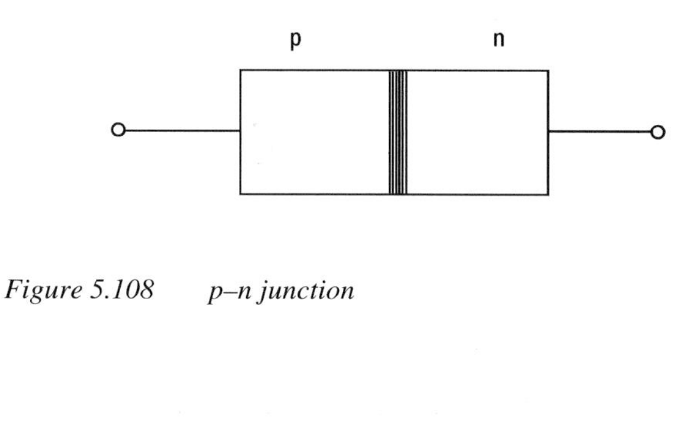


*Figure 5.108 p–n junction*

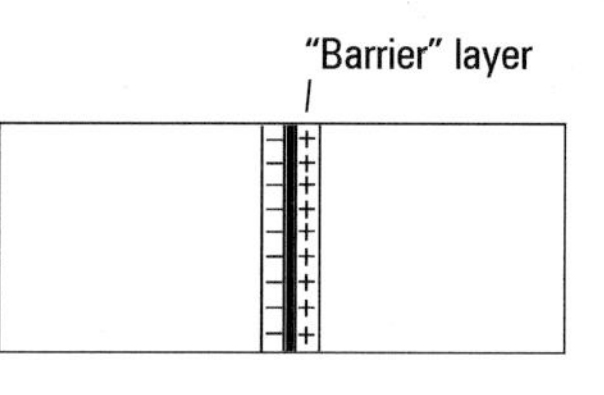

*Figure 5.109 "Barrier" layer*

***Remember***

**"p" has positive charge carriers (holes)**
**"n" has negative charge carriers (electrons)**

# Diodes

If a d.c. supply is connected across the semiconductor diode as in Figure 5.110 the effect is to increase the "barrier" between the n-type and p-type materials (Figure 5.111). In the region of the junction, there are no electrons in the p-type to be attracted across the junction to the more positively charged n-type. Similarly there are no holes in the n-type so no current flows.

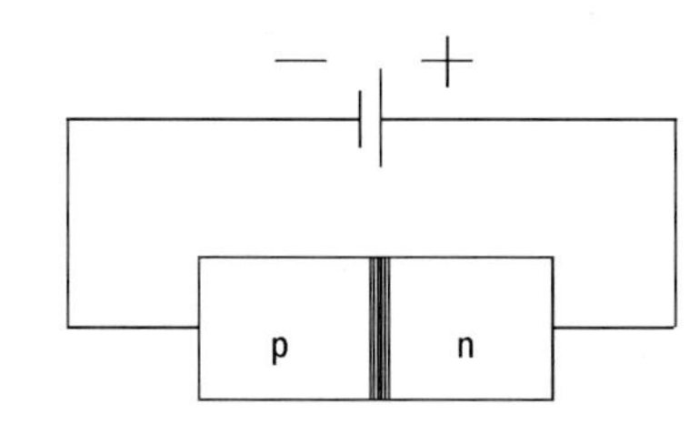


*Figure 5.110* *Reverse bias*

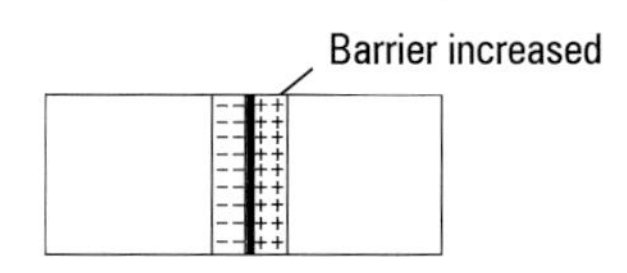


*Figure 5.111*

If the d.c. supply is now reversed, as in Figure 5.112, electrons in the n-type are attracted across the junction and holes in the p-type move (in effect) in the opposite direction.

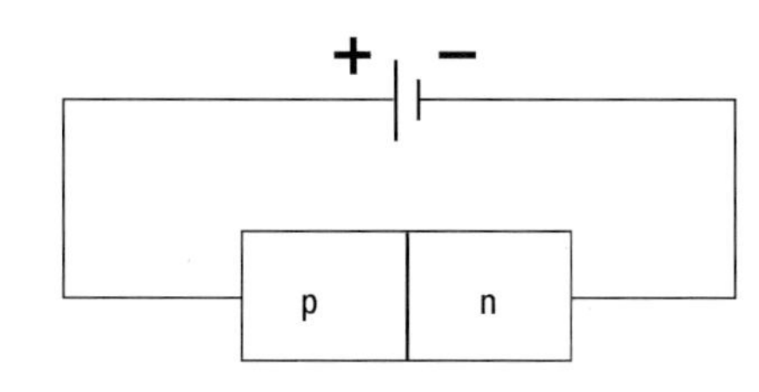


*Figure 5.112*

When a single diode is connected to an a.c. supply (Figure 5.114) it will only allow the current to flow when the p-type is more positive than the n-type. This means that only one half of the current will be conducted. It is in effect switching off the other half of the waveform.

This is known as half-wave rectification.

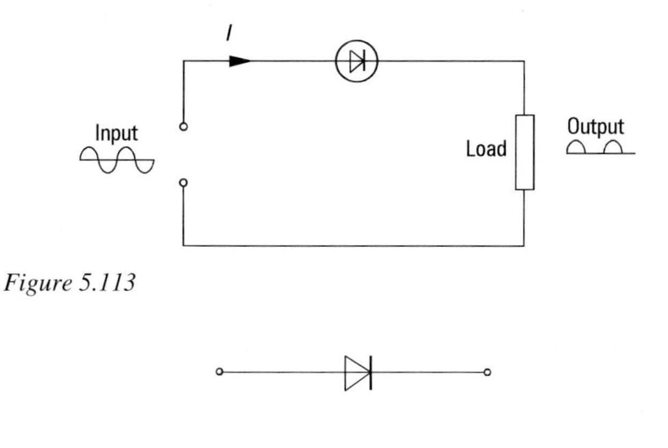


*Figure 5.113*

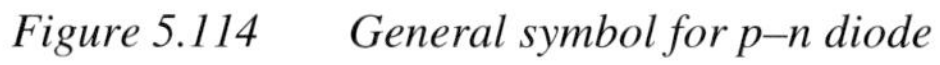

*Figure 5.114* *General symbol for p–n diode*

It is important that diodes are correctly installed and for this purpose it is necessary to ensure that the leads or terminals have been correctly identified (Figures 5.115–5.119).

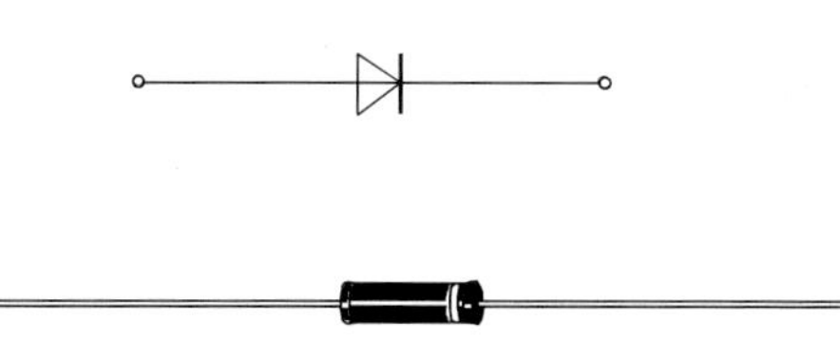

*Figure 5.115* *Indentation to indicate direction of current flow*

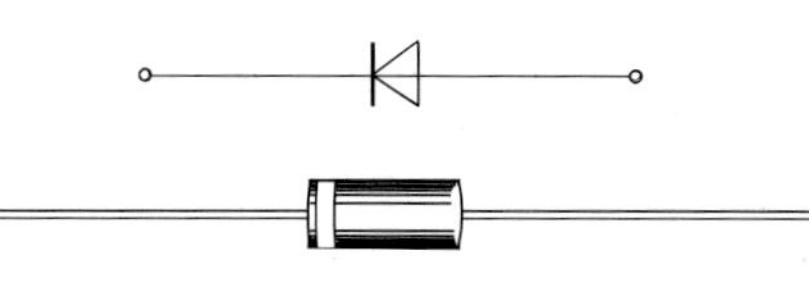

*Figure 5.116* *Band indicates direction of current flow*

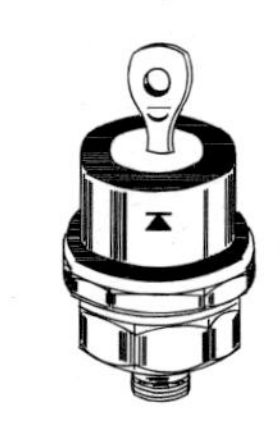

*Figure 5.117* *Stud-mounted diode*

*Figure 5.118* *Surface mount diodes*

Some small plastic encapsulated diodes have one end slightly pointed; this indicates the direction of current flow in normal circumstances, i.e. anode to cathode. If neither end is pointed there may be a slight indentation (as in Figure 5.115) or a different coloured band (Figure 5.116) to indicate which end is the cathode. If there is any doubt then manufacturers' information should be checked for anode/cathode identification.

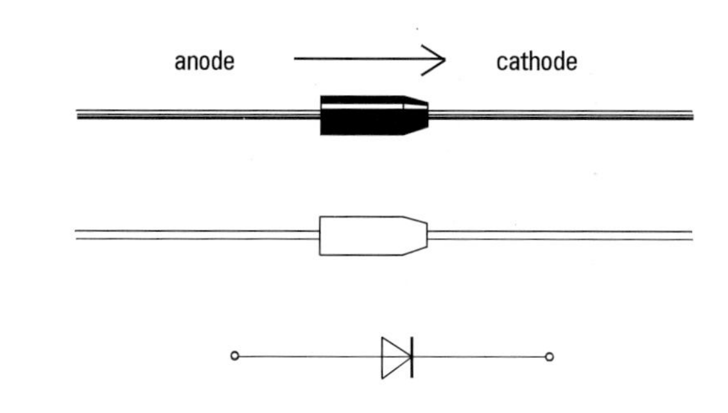


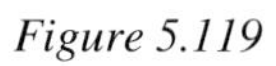

*Figure 5.119*

Full-wave rectifier units frequently make use of encapsulated bridge rectifiers (Figures 5.120 and 5.121), which incorporate four diodes in a bridge configuration. These have four terminals, two a.c. input and two d.c. output, clearly marked + and –.

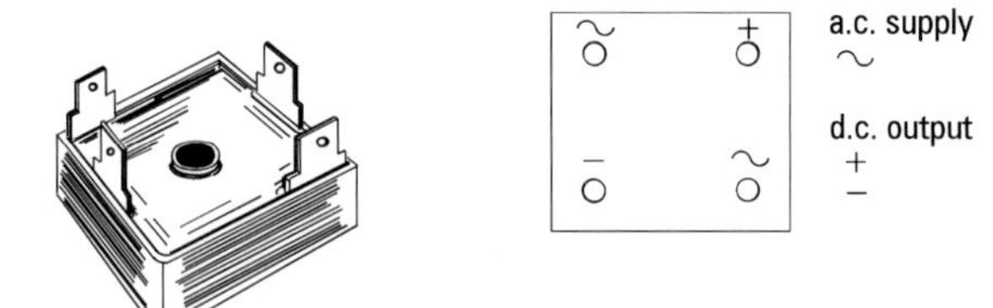


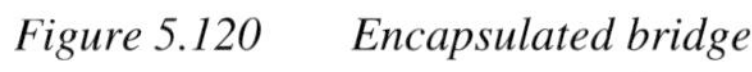

*Figure 5.120* *Encapsulated bridge*

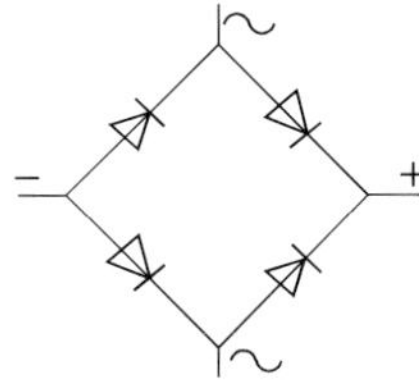

*Figure 5.121* *Full-wave bridge circuit diagram*

# Diodes in common use

## Power diodes

These are diodes in general use which carry currents of 1 A and higher.

## Signal diodes

Signal diodes are small in size and have limited power-handling capability. They are used in high-speed switching circuits or high-frequency communications circuits.

## Zener diode

An ordinary semiconductor diode will only withstand a certain amount of reversed voltage before it breaks down and current flows. Once it has broken down it is permanently damaged. A zener diode is designed to break down without damage at a predetermined voltage and is usually used as a voltage reference. It can be used to provide a constant voltage even when the supply voltage and load current may vary. In Figure 5.122 a 3 V load is supplied even though the 6 V supply may fluctuate slightly. So that the zener diode is not damaged a current limiting resistor (R) is used (Figures 5.122 and 5.123).

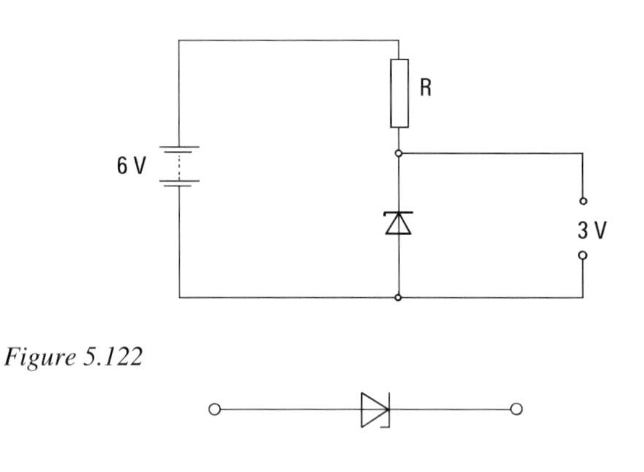


*Figure 5.122*

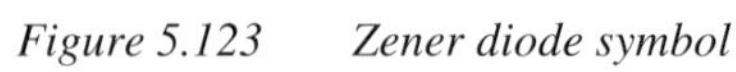

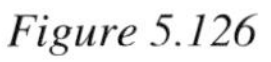

*Figure 5.123* *Zener diode symbol*

## Photodiode

A photodiode (Figure 5.124) is basically a p–n junction which is enclosed in a can with a window. When light shines on the junction, electrons and holes are produced and current can flow until the light is cut off. This makes them suitable for counting and monitoring processes.

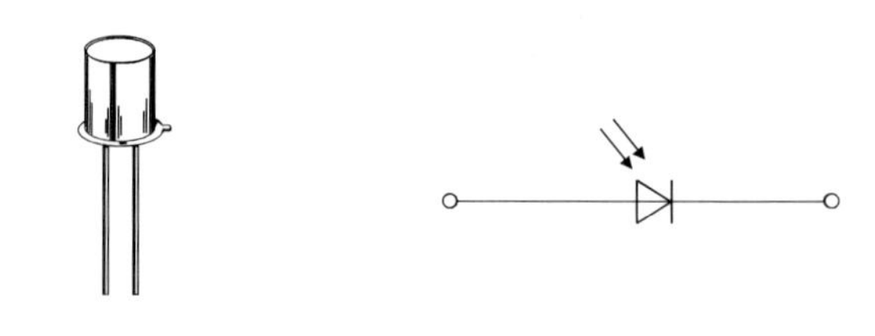

*Figure 5.124* *Photodiode and symbol*

## Light-emitting diode (LED)

This is basically another junction diode, but when it is connected in the forward direction light is given off. The light comes from the junction itself, which is placed close to the surface of the diode and in many cases has a lens built above it to give the maximum output. On single units the cathode is usually indicated by a flat on the body and a shorter lead (Figure 5.125).

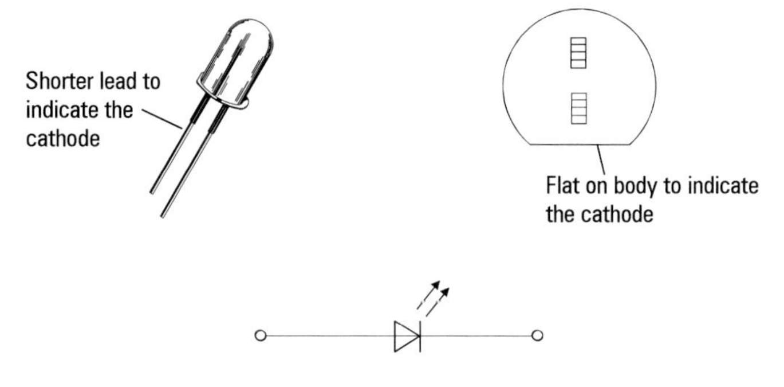


*Figure 5.125* *LED and symbol*

Light-emitting diodes are generally available in red, green and yellow as single or combined units. It is also possible to find blue LEDs or infrared emitters.

When connected in the reverse direction there is no light emitted.

LEDs can be packaged with several in the same unit. When these are used the manufacturers' details need to be studied. An example of a typical arrangement for a four-LED encapsulated unit is shown in Figure 5.126.

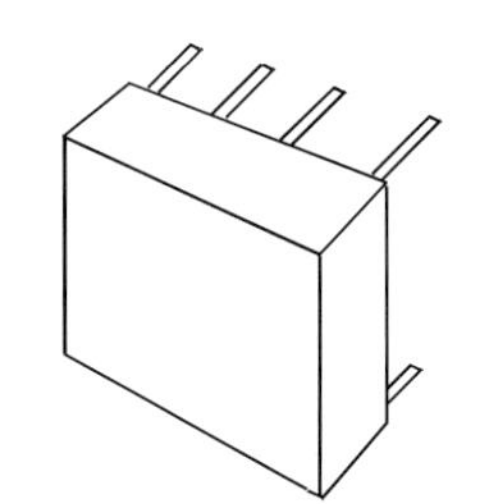

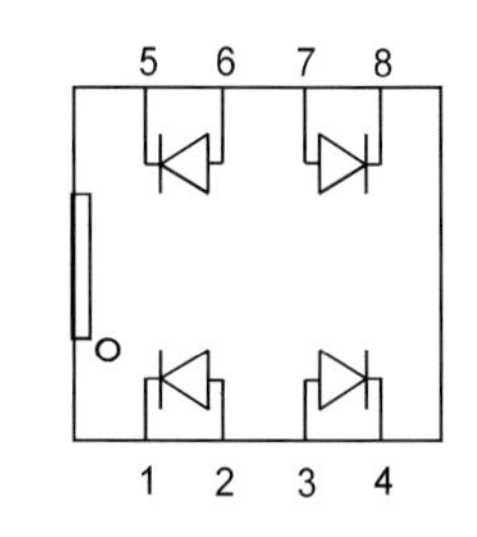


*Figure 5.126*

Other configurations are found, such as seven-segment LEDs (Figure 5.127) or alphanumeric displays.

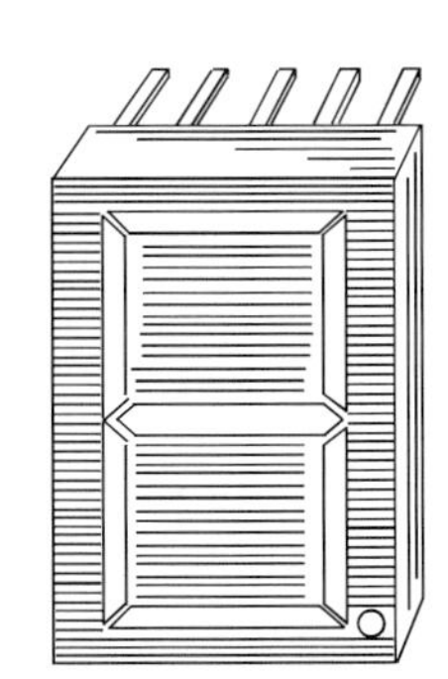

*Figure 5.127* *Seven-segment LED*

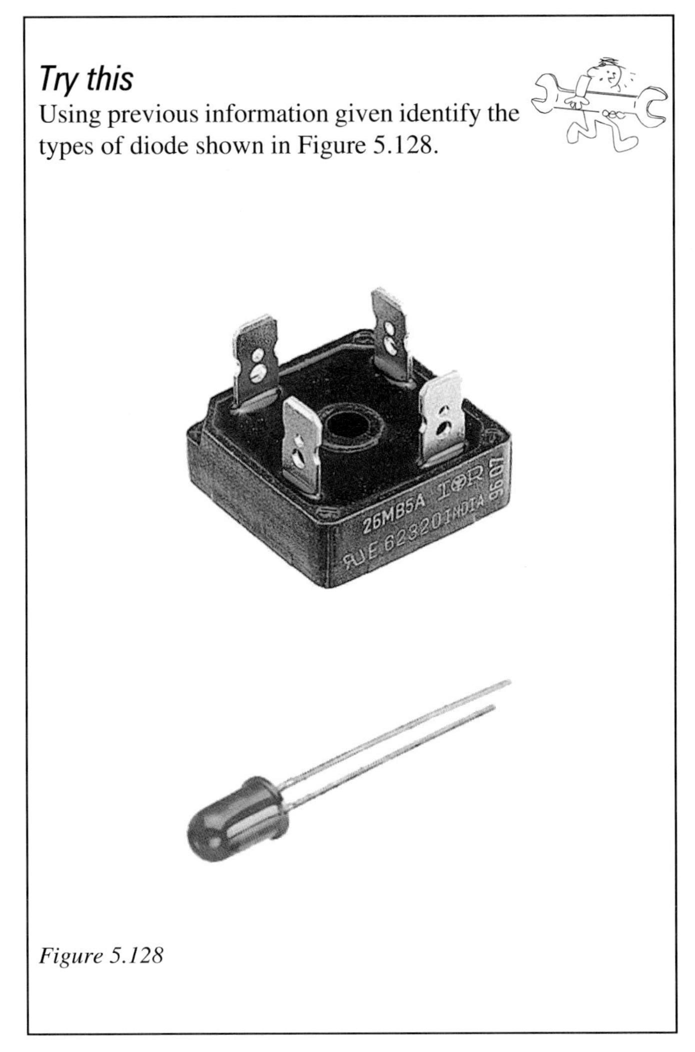

## *Try this*

Using previous information given identify the types of diode shown in Figure 5.128.

*Figure 5.128*

# Transistors

The transistor is a three-terminal device which is capable of amplifying current. A single bipolar transistor has three connections: "emitter", "base" and "collector". The bipolar transistor can be used as a device where a signal current of relatively small proportions can be fed into the base, this will produce a larger current in the collector circuit which can be used for a higher power application. The collector current will only flow as long as the base current is present.

Figure 5.129 shows a circuit where a small input signal controls the operating coil of a relay.

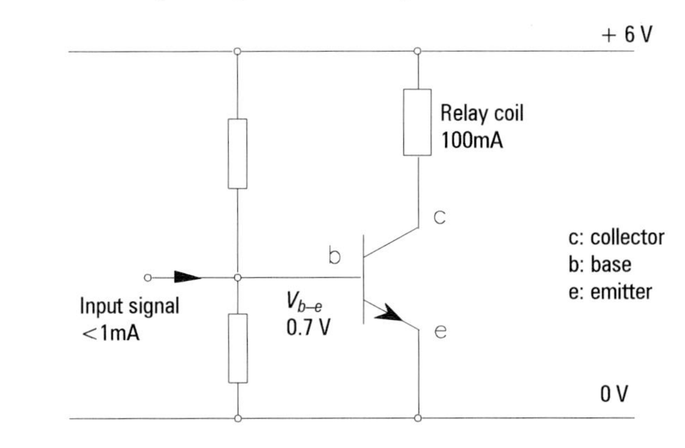


*Figure 5.129*

The identification of the three connections depends on the type of enclosure and mounting arrangement used. Figure 5.130 shows some examples, but the exact configuration for a particular transistor should always be checked with the manufacturer's data before it is connected into the circuit.

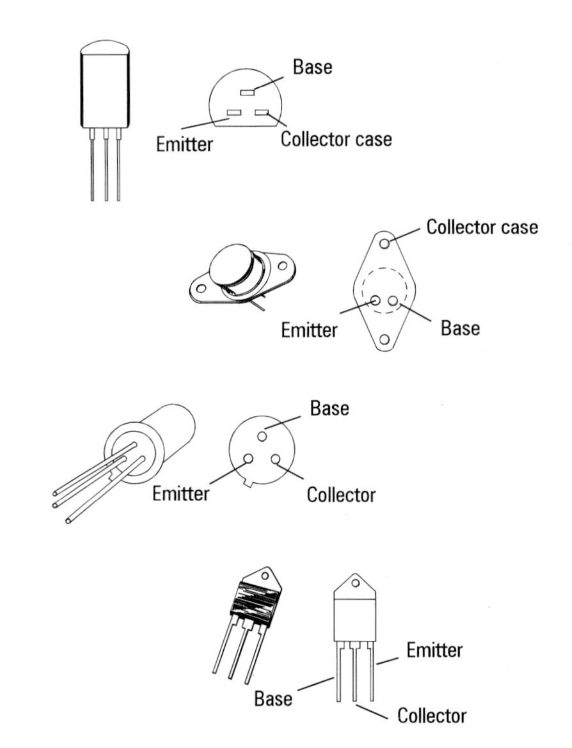


*Figure 5.130* *Pin connections vary with type*

There are two main types of transistor: pnp and npn (Figures 5.131 and 5.132).

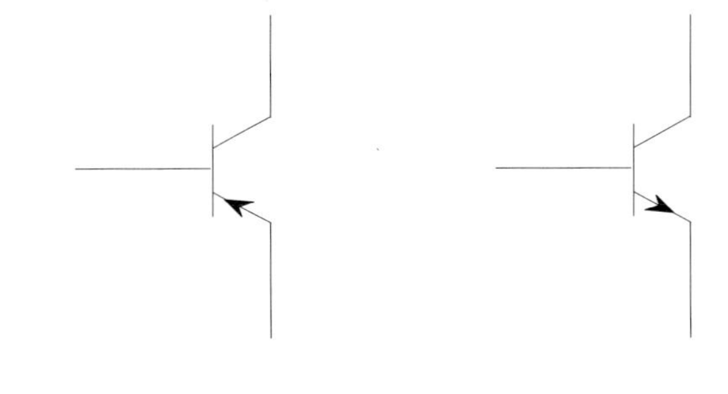

*Figure 5.131* *pnp symbol* *Figure 5.132* *npn symbol*

## Field-effect transistor (FET)

The FET (Figures 5.133 and 5.134) has three terminals, called the drain, the gate and the source. These are equivalent to the collector, base and emitter (respectively) of a bipolar transistor.

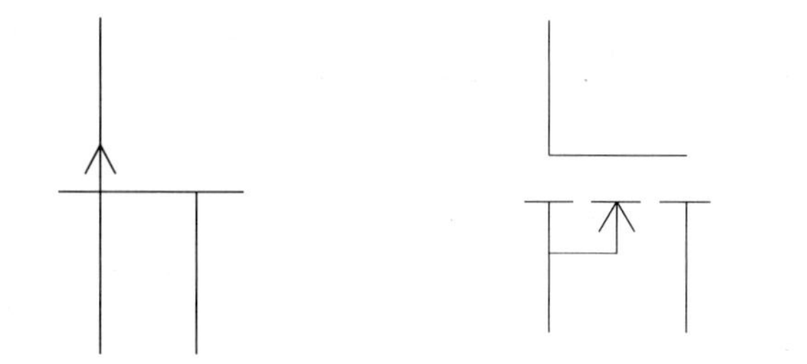

*Figure 5.133 JFET symbol* *Figure 5.134 IGFET symbol*

## Thyristor

A thyristor (Figure 5.135) is a type of silicon-controlled rectifier (SCR) which has cathode, anode and gate connections. The gate connection is usually smaller than the other two and the anode connection is often a stud fixing to a heat sink. A **heat sink** dissipates any heat generated within the device when it is handling high currents.

It may be necessary to replace a failed thyristor or other device which is mounted on a heat sink. It is important, when refitting the device to the heat sink, to replace the thermal compound which provides the heat transfer between the semiconductor and the sink. Any mounting isolating washers should also be replaced in the correct positions.

*Figure 5.135 Thyristor and symbol*

Thyristors are used in switching circuits. They are four-layer devices which do not conduct in either direction until a short pulse is applied to the gate. They will continue to conduct after the gate pulse has been removed (Figure 5.136).

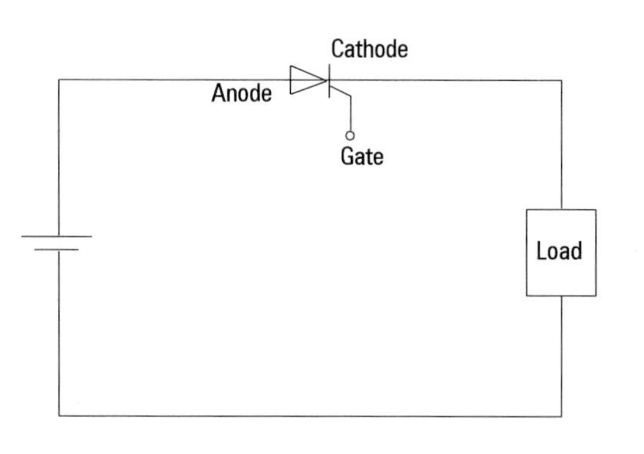


*Figure 5.136*

The thyristor is a d.c. device and will only conduct in the direction indicated. If connected to an a.c. supply it will cease conducting after the supply current has reached zero and will only switch on again if the gate pulse is reintroduced in the next or some subsequent positive half-cycle.

The circuit shown in Figure 5.137 includes two thyristors, SCR1 and SCR2, connected "back-to-back". This then allows for connection to a.c. circuits so that one of the thyristors will operate on each half cycle.

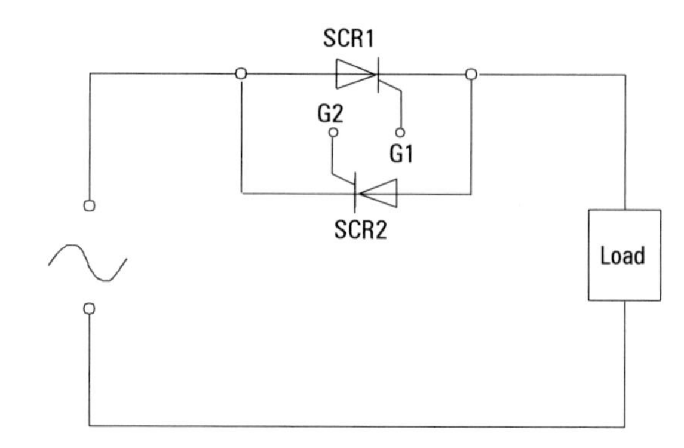


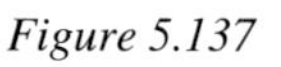

*Figure 5.137*

## Triac

A triac is a bi-directional switching device which can be used in a.c. circuits. Triacs are found in motor speed control circuits and light dimmers. Their case and connection layout often get them confused with transistors, as they can look very similar. Triacs are, however, equivalent to a pair of thyristors in a single case. A typical configuration is shown in Figure 5.138.

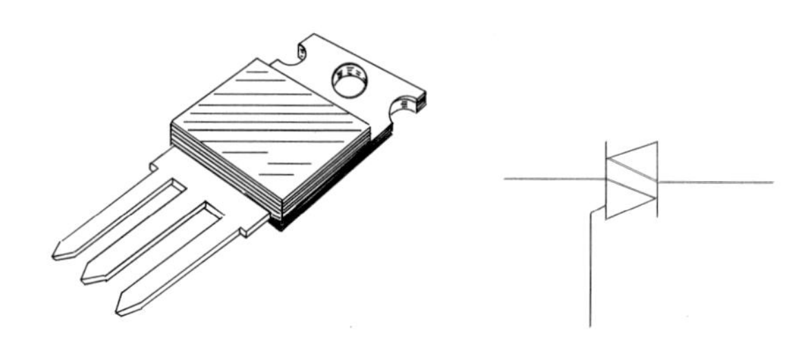

*Figure 5.138 Triac and symbol*

## Diac

A diac (Figure 5.139) is a two-terminal device which conducts in either direction, but only after the voltage across them has reached the level required for conduction to take place. It has only two connections and can be confused with a diode if not careful.

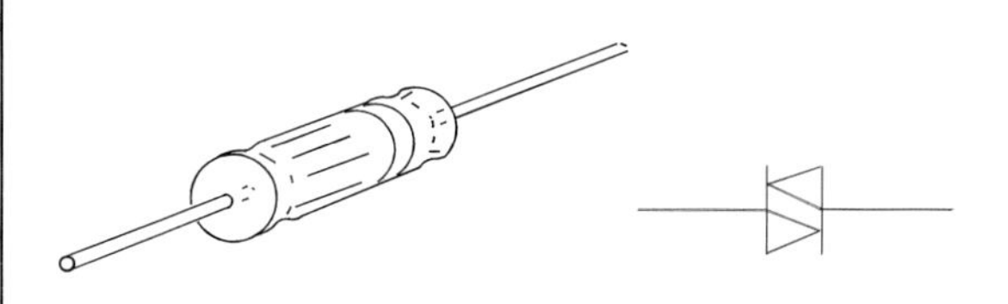

*Figure 5.139 Diac and symbol*

# Integrated circuits

The integrated circuit (IC) is a complete electronic circuit formed from one piece of semiconductor material. The active component may consist of several thousand (or even million) transistors along with their associated equivalent resistors, diodes and capacitors all incorporated in a tiny chip of silicon and encased in a plastic package. The range and packaging of devices is continually increasing and developing to cope with the ever-demanding market. It is recommended that you refer to current manufacturers' catalogues for the latest information.

Integrated circuits are generally grouped into two classes categorised by the types of input/output signals. Voltages that can vary over the range of the power supplies are called analogue or linear devices. Voltages that have only two voltages are digital circuits.

An example of an analogue/linear integrated circuit is an operational amplifier of the 741 series. It is a device which has an open loop gain of many hundreds of thousands which is modified by the connection of external resistors to any value required by the user (say 1000).

An example of a digital integrated circuit is a NAND gate of the 7400 series which we will be looking at in the following chapter.

Packaging exists in many forms. A common package, shown in Figure 5.140, is an eight-pin dual in-line assembly and the pins are numbered from the left of the notch looking down and following anti-clockwise round the component.

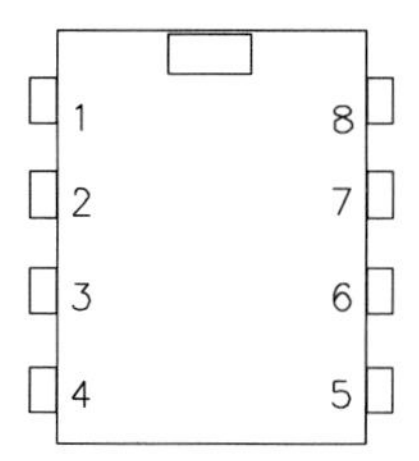


*Figure 5.140*

A 14 pin dual in-line package may have a dot next to pin No. 1 and the following pins numbered as before (Figure 5.141).

*Figure 5.141*

No matter how many pins a DIL package has, and it may be as many as forty plus, the numbering system is the same.

## Printed circuit boards

A printed circuit board (PCB; Figure 5.142) is an insulated base with tracks of conducting material running in an intricate pattern over one or both sides of the board. There are also multi-layer boards with possibly up to as many as 10 layers.

The board may start life as a sheet of copper-faced laminate from which the unwanted parts are usually removed by chemical etching.

The main advantage of the printed circuit board is that once the design is complete and a master model has been made, an infinite number of copies can be produced from the same pattern.

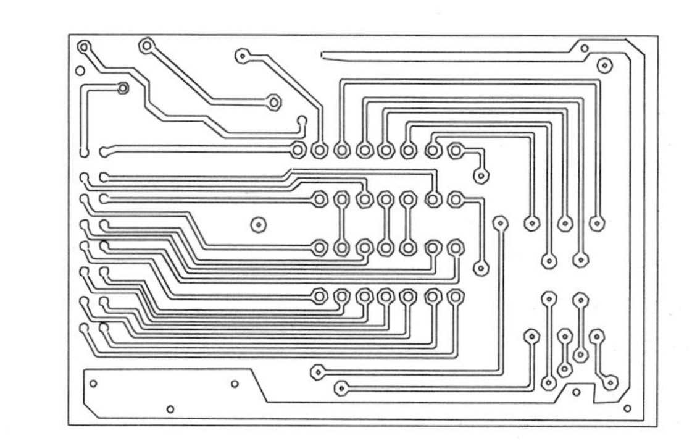

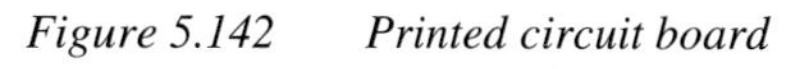

*Figure 5.142 Printed circuit board*

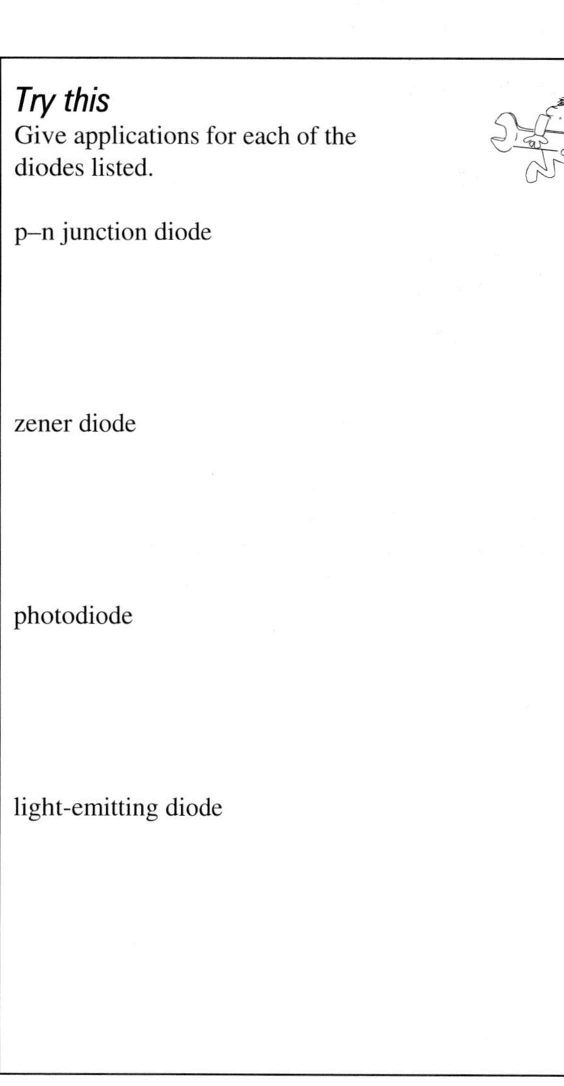

### *Try this*

Give applications for each of the diodes listed.

p–n junction diode

zener diode

photodiode

light-emitting diode

## *Points to remember*

Semiconductor materials are, as the name implies, somewhere between insulating and conducting materials. By doping semiconductor materials it is possible to create new ones that respond to positive or negative charges. These materials are usually called "____" and "____" types.

By connecting these materials in different ways within components it is possible to create semiconductor devices such as diodes, __________________, __________________ and __________________

A diode is used in a d.c. circuit as a switch and in an a.c. circuit as a ______________

A thyristor is a type of ____________ ______________ rectifier.

What do the following abbreviations stand for?

FET

LED

IC

PCB

SCR

## *Self-assessment multi-choice questions*

**Circle the correct answer in the grid below.**

1. The "p" in a p–n junction represents a
   (a) protective material
   (b) powered junction
   (c) material with positive charge carriers
   (d) polarity changed junction
2. The general symbol for a p–n diode is
   (a) (b)

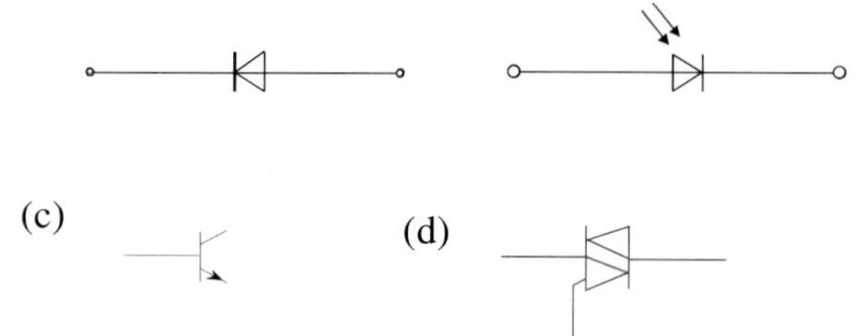

   (c) (d)

3. An example of a silicon-controlled rectifier is a
   (a) diode
   (b) full-wave bridge
   (c) transistor
   (d) thyristor
4. The electronic component shown in Figure 5.143 is a
   (a) transistor
   (b) thyristor
   (c) triac
   (d) light-emitting diode

*Figure 5.143*

5. The graphical symbol shown in Figure 5.144 represents
   (a) a transistor
   (b) a triac
   (c) a diode
   (d) a thyristor

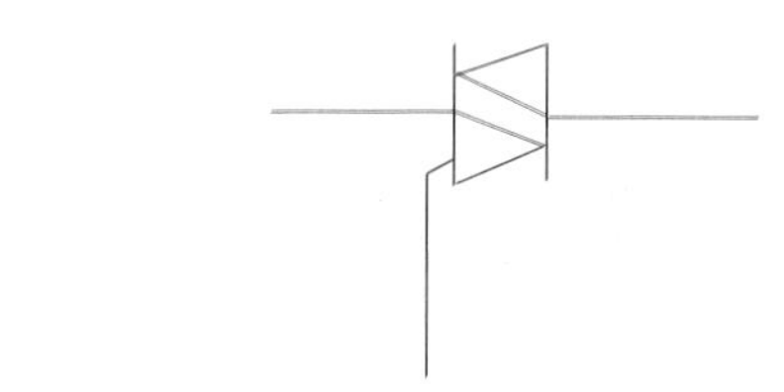

*Figure 5.144*

### *Answer grid*

| | | | | |
|---|---|---|---|---|
| 1 | a | b | c | d |
| 2 | a | b | c | d |
| 3 | a | b | c | d |
| 4 | a | b | c | d |
| 5 | a | b | c | d |

# 6

# Electronic Symbols and Diagrams

Complete the following to remind yourself of some important facts from the previous chapter.

What is commonly referred to as a "pot"?

Where would a thermistor be used?

Resistors can usually be recognised by their shape and the presence of ________________ or ________________
What is capacitance usually measured in?

The type of capacitor often used for radio tuning is a________ ______________________ capacitor.
Transformers used on radio frequency usually have adjustable ferrite cores and are usually enclosed in a______ to screen them from other parts of the circuit.
A diode is a device which will permit current flow in one direction. The cathode end of a diode can be identified by
______________________

A transistor is a device capable of amplifying current and has __________ terminals. These connections are called:

The two main types of transistor that we noted in the previous chapter are:

**On completion of this chapter you should be able to:**

- ◆ recognise the need for circuit, block, layout and logic diagrams
- ◆ recognise the need to use standard symbols
- ◆ relate some electronic graphical symbols to their components
- ◆ draw block and circuit diagrams from layout diagrams
- ◆ draw block and layout diagrams from circuit diagrams
- ◆ describe the action of basic two-input logic gates
- ◆ complete the revision exercise at the beginning of the next chapter

## Use of diagrams

With electronics becoming more and more complex it is important to have diagrams that are clear and understandable. To help to achieve this there are diagrams which are designed to show different aspects. These diagrams are

- block
- circuit
- layout
- logic

**Block diagrams** are used to show a sequence of actions or the interaction between defined sections. They show how parts of a design interconnect, without showing too much detail.

**Circuit diagrams** show, in detail, which component pins are connected together, with little information as to how this is achieved. They allow the operation of the circuit to be understood. The components are shown using standard graphical symbols.

**Layout diagrams** give a diagrammatic picture of how the finished circuit will appear. The conductors and components are represented by scaled shapes of the actual objects.

A **logic diagram** shows how logic gates or other digital circuits can be combined to provide a control function which will deliver one or more output signals when certain input conditions are present.

Other types of diagram you may come across are the flow diagram, which will show a sequence of events, or a timing diagram or chart, which shows time relationships between the different signals of digital computing components.

The previous chapter in this book examined resistors, capacitors, inductors and semiconductors. Each component shown has a British Standards Institution symbol to represent it on drawings. Note that you may find elsewhere that some symbols are shown with a circular envelope on semiconductor devices. The following four pages give a summary of this information, together with photographs of the components concerned.

# Resistors

## Symbols

general

potentiometer

preset

light-dependent resistor

thermistor

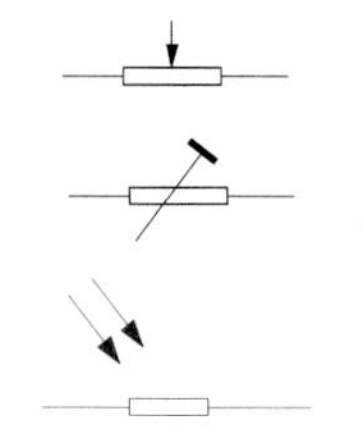

*0.5 W metal film resistor*

*Surface mount*

*Vireous enamel wirewound resistor*

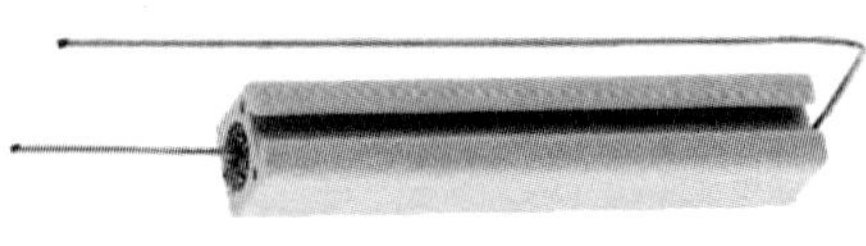

*Ceramic body wirewound resistor*

*Aluminium-housed wirewound resistor (mounted on a heatsink)*

*DIL resistor package*

*SOIC resistor network*

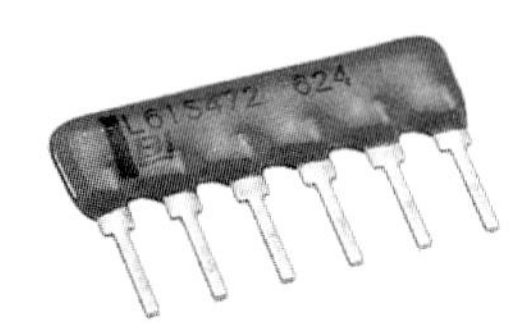

*SIL resistor package*

# Inductors

## Symbols

iron cored

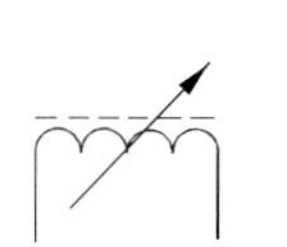

ferrite or iron dust cored

variable with ferrite or
iron dust cored

*RF inductors with ferrite cores*

*IF coil with adjustable ferrite core*

*Low-frequency inductor*

# Capacitors

## Symbols

general

polarised

variable

trimmer

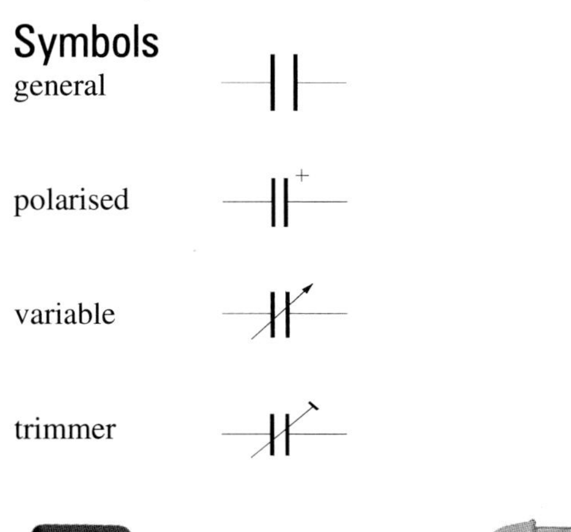

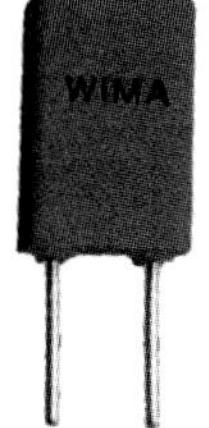


Polypropylene film capacitor

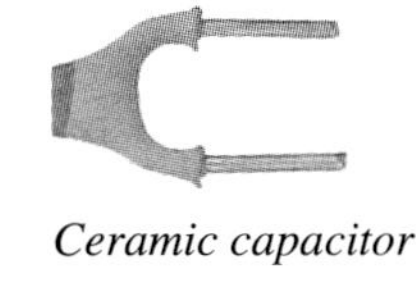

*Ceramic capacitor*

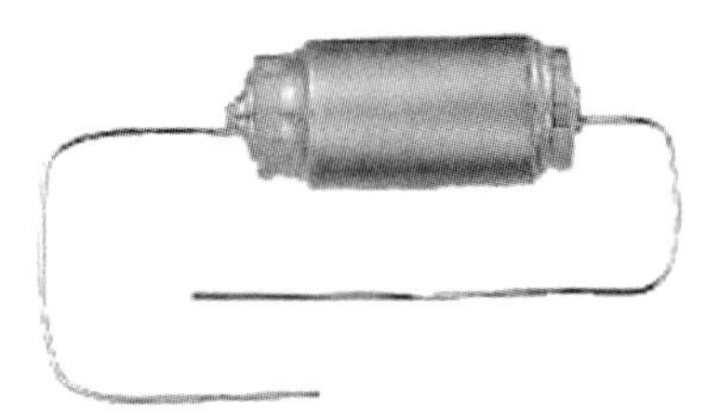

*Polystyrene capacitor*

*Polypropylene capacitor*

*Electrolytic capacitor*

*Surface mount aluminium electrolytic capacitor*

*Tantalum bead capacitor*

*Aluminium electrolytic capacitor*

*Variable capacitor*

*Variable capacitor*

# Transformers

## Symbols

iron core

ferrite core

autotransformer

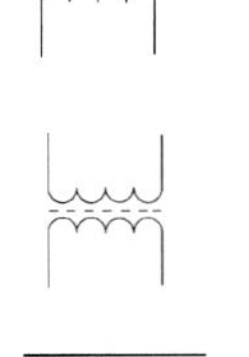

*PCB mounted transformer*

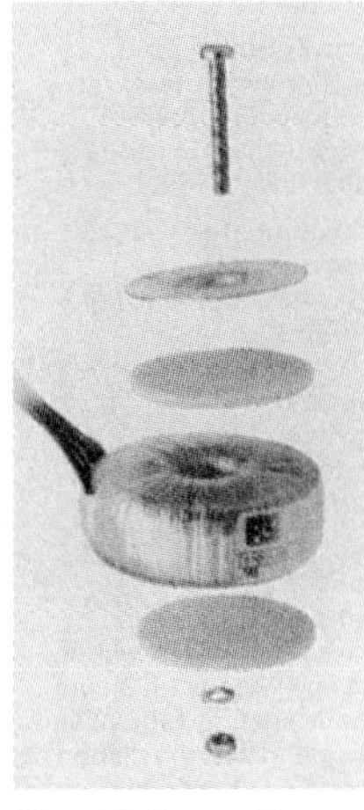

*Toroidal transformer*

*AF transformer*

# Semiconductor diodes

## Symbols

general

zener

light sensitive

light emitting

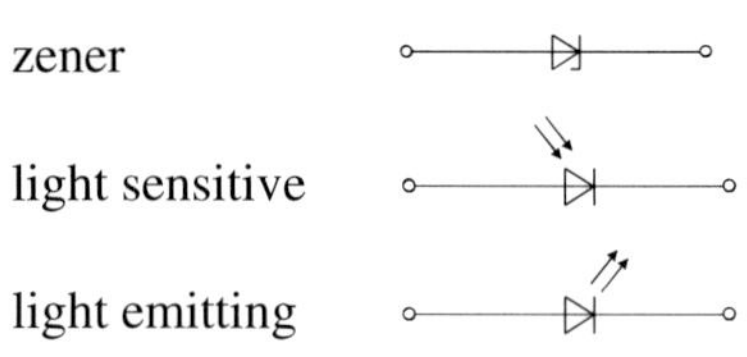

*Surface mount diode*

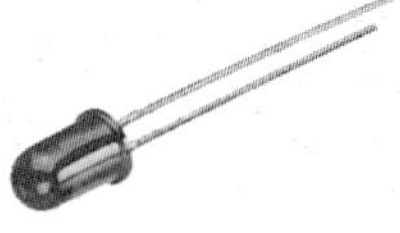

*Standard LEDs*

# Transistors

## Symbols

npn

pnp

FET

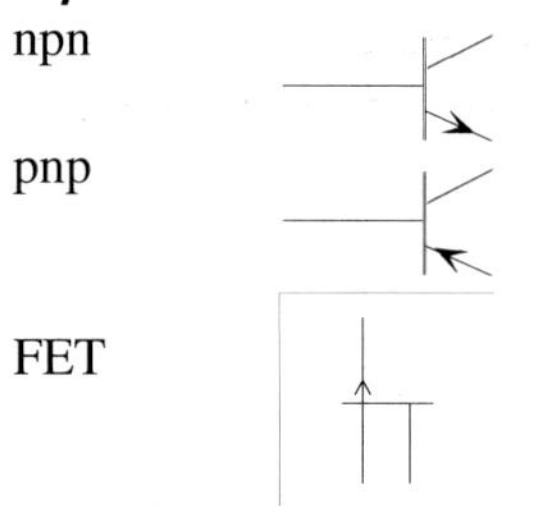

# Control rectifiers

## Symbols

thyristor

triac

diac

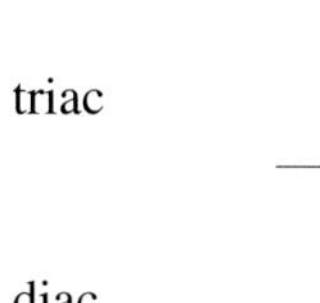

*Thyristors*

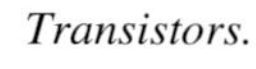

*Diac*

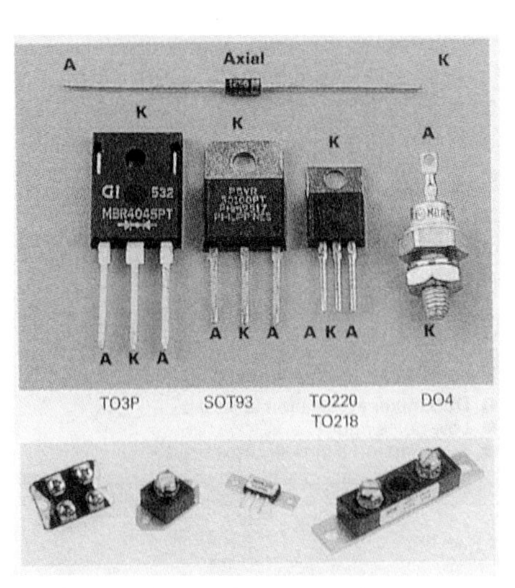


*Diodes (numbers refer to type of packaging)*

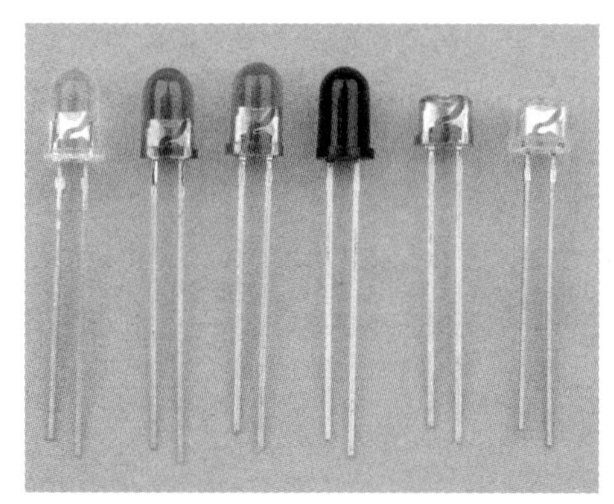

*Photodiodes and phototransistors*

*Photodiode*

*Seven-segment LED display*

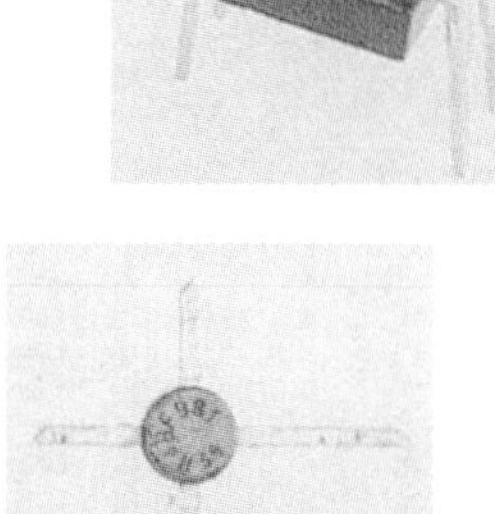

*Transistors.*

*Note: SCRs, transistors and even diodes have common packaging. Some components can only be identified from the manufacturer's catalogue numbers.*

# Integrated Circuits

*An example of an IC – further examples can be seen on page 86*

## Others

relays (see below)

switches (see below)

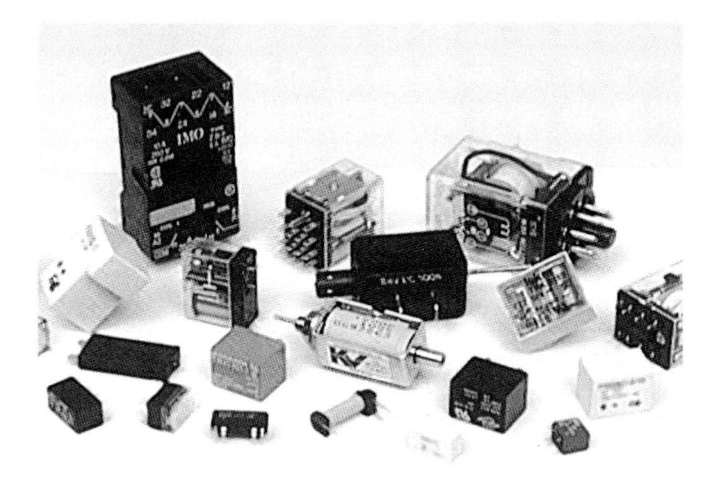

*A selection of relays*

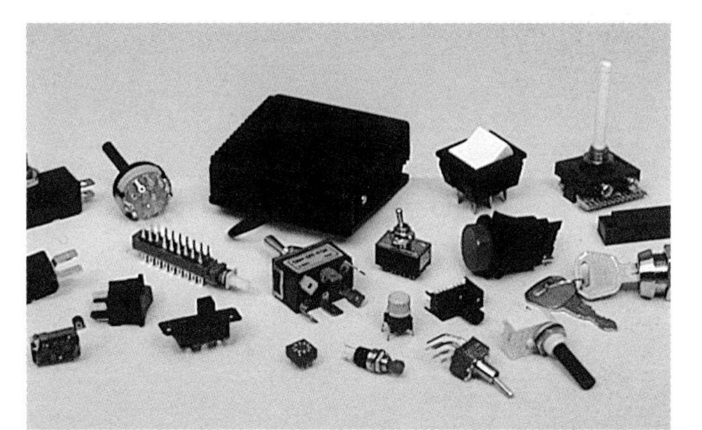

*A selection of switches*

## Block diagram symbols

These are simplified symbols contained within a single block.

Examples of complete assemblies can be shown as:

Rectifier

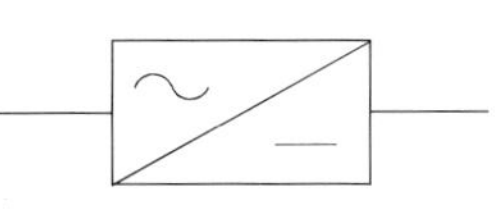

Amplifier

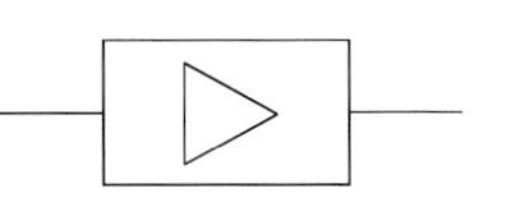

Transformer

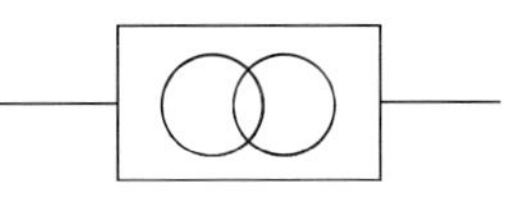

Inverter

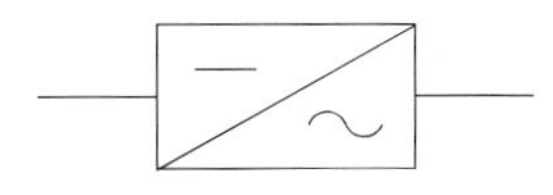

There are many more available in BS 3939.

## Circuit symbols

All of the electrical and electronic graphical symbols in the British Standards Institution 3939 follow a few simple rules.

For example, this is the general symbol for a resistor.

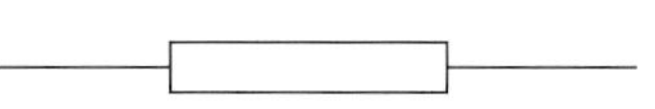

Variable components are shown with an arrow through them.

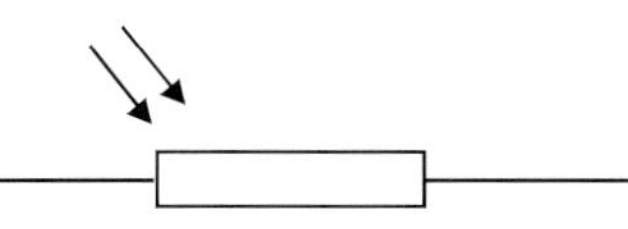

If it is an adjusted preset value the arrow head is replaced with a bar.

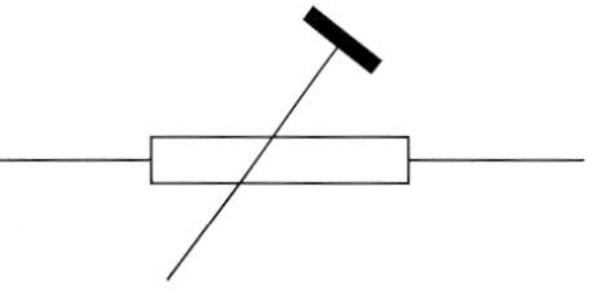

Other examples using additional symbols are:
light-emitting and light-sensitive.

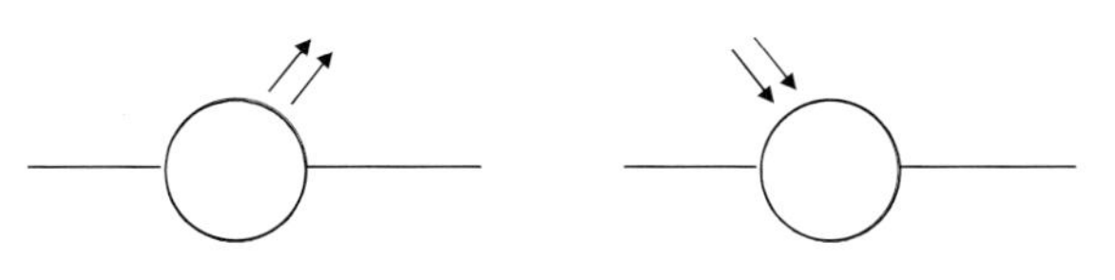

The main component may be a diode or a resistor.

## Examples of the use of diagrams

The circuit used for this example is a light-activated switch which could be used to control a lamp so that it automatically comes on when the light reaches a certain level. The circuit operates a relay coil so that the lamp is separate to the electronic components.

## Block diagram

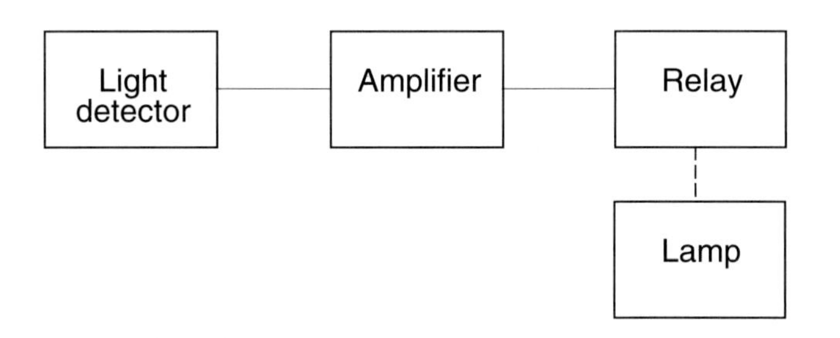


## Circuit diagram

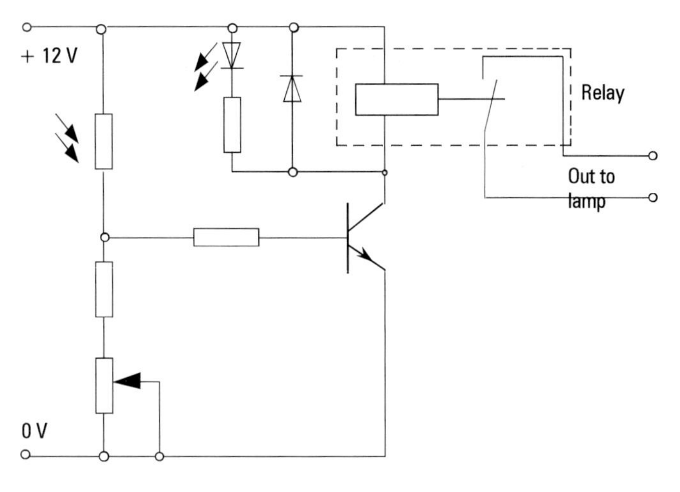

*Figure 6.1*

## Layout diagram

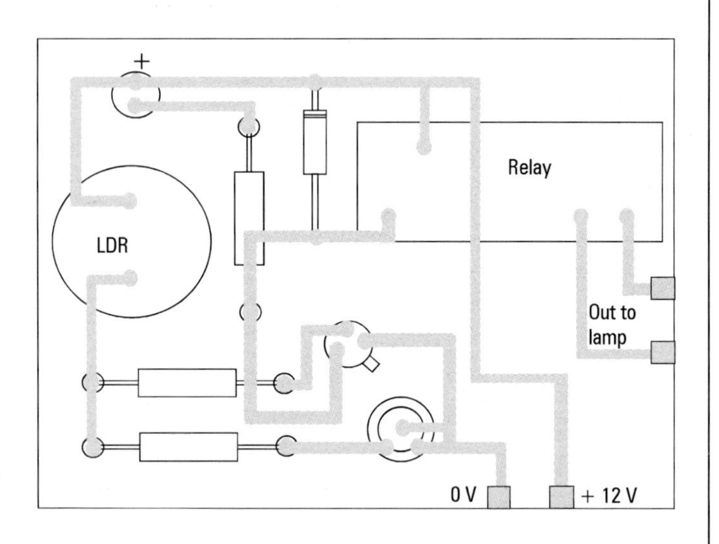


*Figure 6.2*

## Component positional reference systems

As electronic diagrams become more complex it is difficult to include on the drawing all of the relevant information associated with each component. One method of overcoming this problem is the use of component position reference drawings (Figure 6.3). By identifying each component in this way a key can be drawn up giving all of the relevant information.

## Component position reference drawing

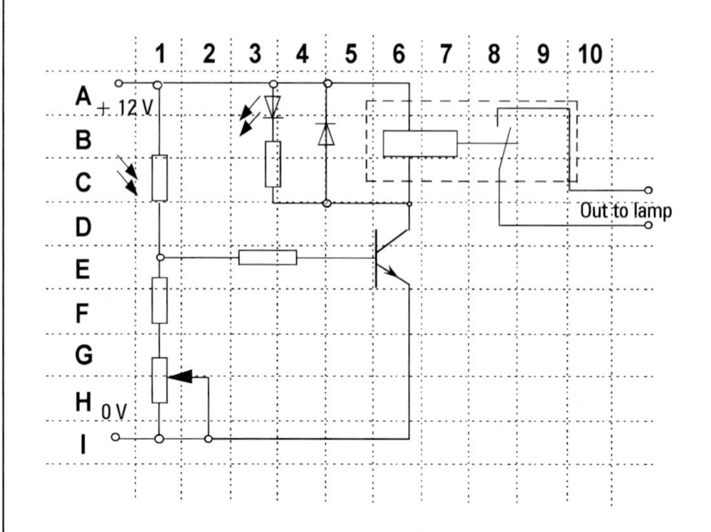


*Figure 6.3*

## Key

| | | |
|---|---|---|
| A3 | light-emitting diode (LED) | |
| B4 | diode | 1N4148 |
| B6 | relay | minimum coil resistance 110 Ω |
| C1 | light-dependent resistor (LDR) | |
| C3 | resistor | 1 kΩ |
| E3 | resistor | 1.5 kΩ |
| E6 | transistor | 2N3053 |
| F1 | resistor | 300 |
| H1 | potentiometer | 5 kΩ |

## Circuit description

A circuit description is often used in maintenance manuals to aid servicing.

The light-activated switch circuit used (Figure 6.4) could be described as follows:

> The small amount of current change in the detector circuit is amplified by the transistor so that it is sufficient to operate the relay.
>
> The light-dependent resistor (C1) varies from about 400 Ω in sunlight to 1 MΩ in the dark, and this, with the series combination resistor (F1) and potentiometer

(H1) makes a potential divider to feed a voltage to the current limiter resistor (E3). The voltage at the base transistor (E6) then rises above the cut-off voltage and the transistor switches on and current flows from the collector to the emitter. This current flows through the relay coil (B6) causing a magnetic field to force the "spring" biased contacts to close. The current through the transistor also provides an LED (A3) to be illuminated as an indicator and the resistor (C3) limits the current to a safe level. When the external light level is extinguished the voltage at the transistor drops below its base/emitter threshold to conduct and switches the transistor off. The magnetic field cuts off and produces a potential voltage swing above the supply voltage due to its self-inductance. This potential positive voltage (measured above the supply voltage) rises. At about 0.7 V diode (B4) switches on, allowing current to flow and hence preventing a further voltage rise.

In the circuit used in this example the relay contacts could be extended to include a light dimmer, as shown in Figure 6.5. The circuit description could then continue in the following way:

A light dimmer functions by switching its a.c. current on only for part of each cycle. Consider that initially the relay contacts are closed and the instantaneous voltage at the a.c. supply is at zero. As this voltage increases so will the voltage across the capacitor (G14), but this is delayed by the time taken for the capacitor to charge. When this reaches about 30 V the diac (E15) conducts, which in turn switches on the triac (D16). This allows the current to flow and the lamp "lights". This continues until the supply voltage drops to zero when the triac blocks the current and extinguishes the lamp. A similar sequence then occurs on the second, negative, half of the cycle. The trimmer (D14) can be adjusted to set the light output level. The resistor (D18) and the capacitor (G18) fitted across the triac prevent "noise" spikes from causing false triggering.

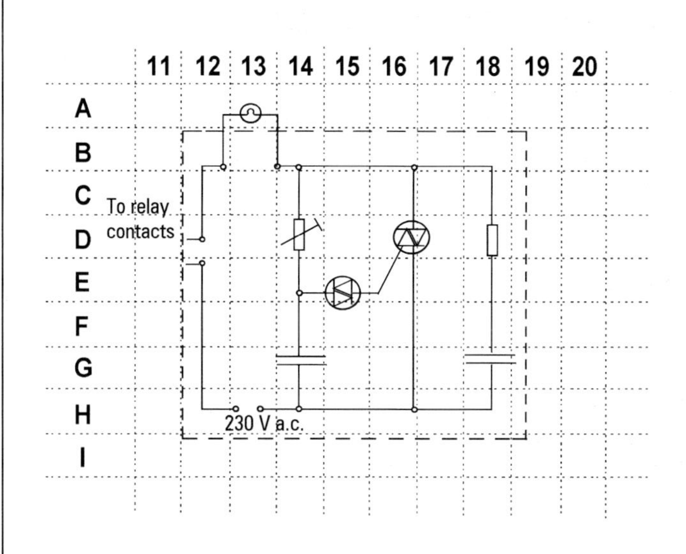


*Figure 6.5* *Component position reference drawing for the circuit containing the light dimmer*

## Key

| | |
|---|---|
| A13 | lamp |
| D14 | trimmer (resistor at a preset value) |
| D16 | triac |
| D18 | resistor |
| E15 | diac |
| G14 | capacitor |
| G18 | capacitor |

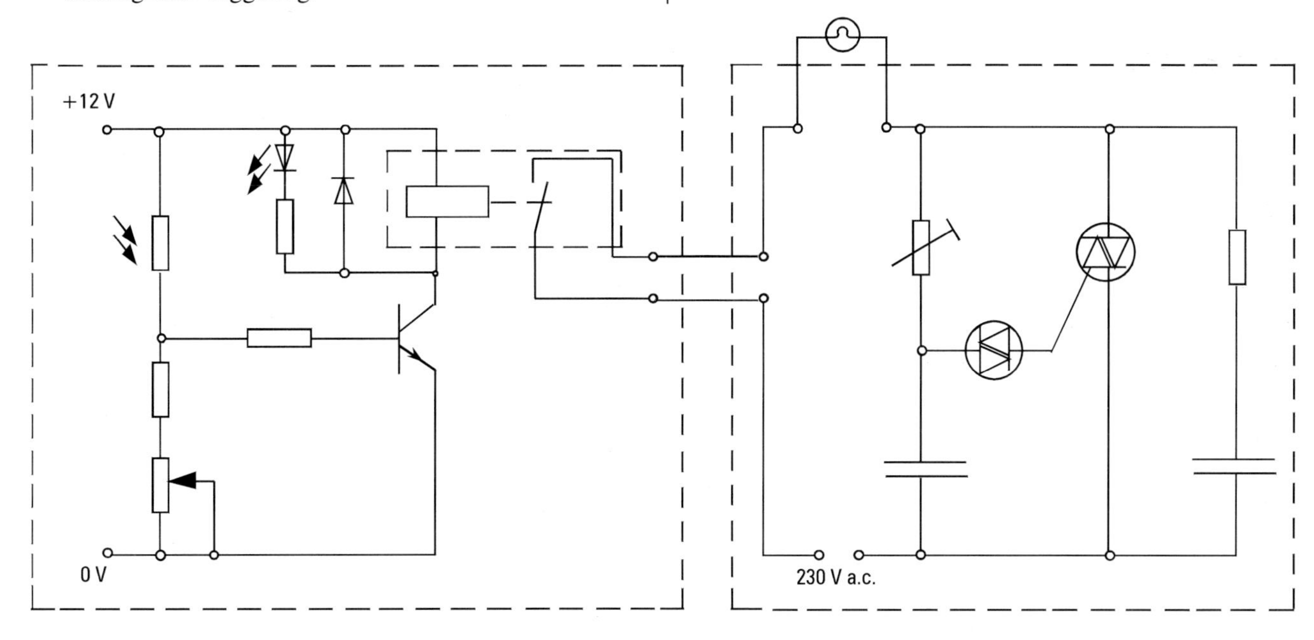


*Figure 6.4* *Complete circuit including light dimmer*

*Example*

## Regulated d.c. voltage supply

So that a voltage supply is constant a voltage regulator is used. In this circuit a transformer is supplied with 240 V a.c. This is transformed down, rectified and then regulated to give the constant voltage output.

### Block diagram

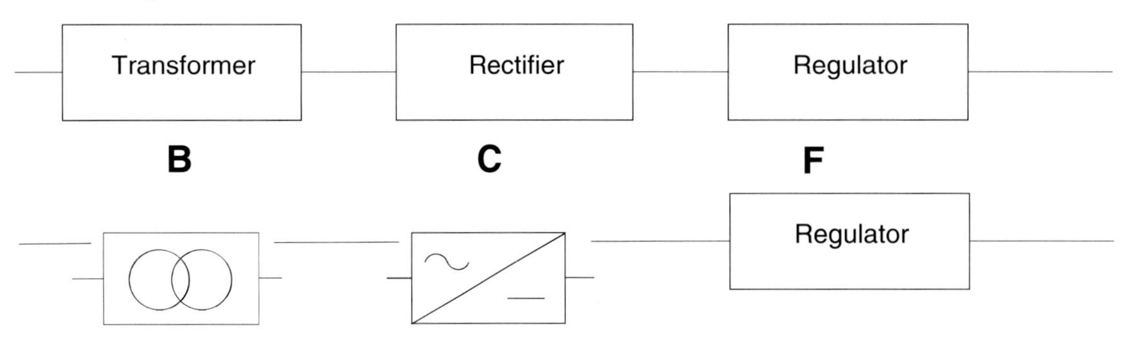


### Circuit diagram

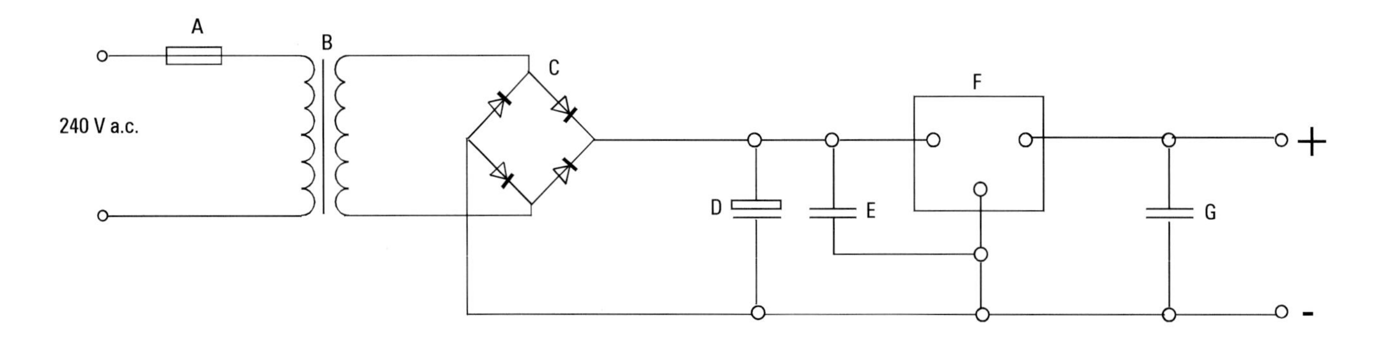


### Layout diagram

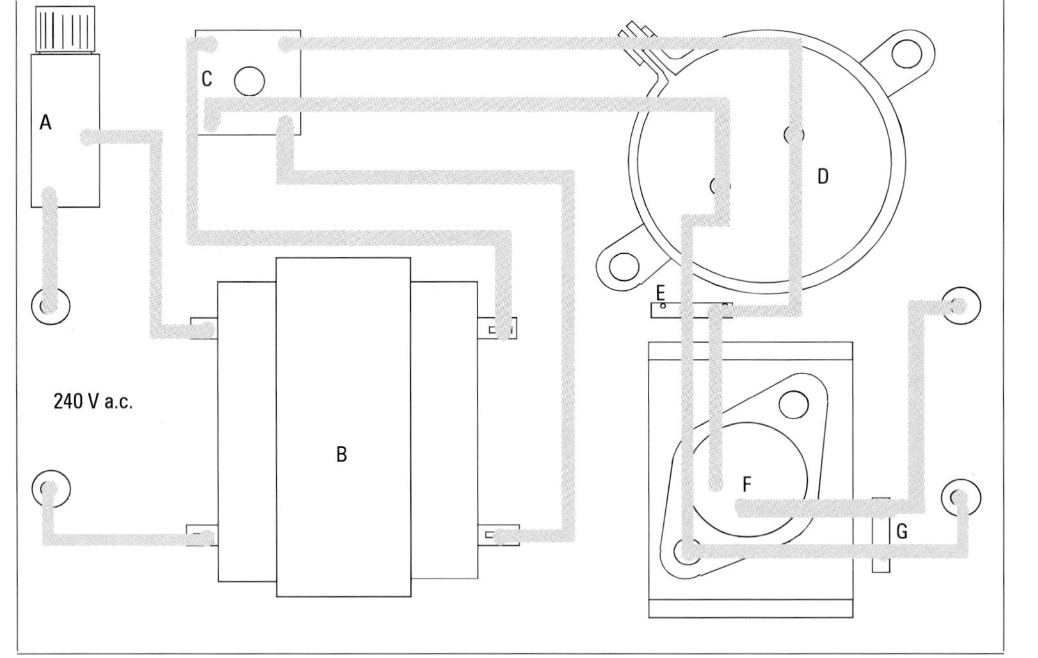


### Parts list

A fuse unit
B 240 V to 12 V double wound transformer
C full-wave rectifier bridge
D smoothing electrolytic capacitor
E capacitor
F regulator
G capacitor

*Try this*

## A.C. rectified supply

Draw the circuit and block diagrams for the layout shown below.

## Block diagram

## Circuit diagram

## Layout diagram

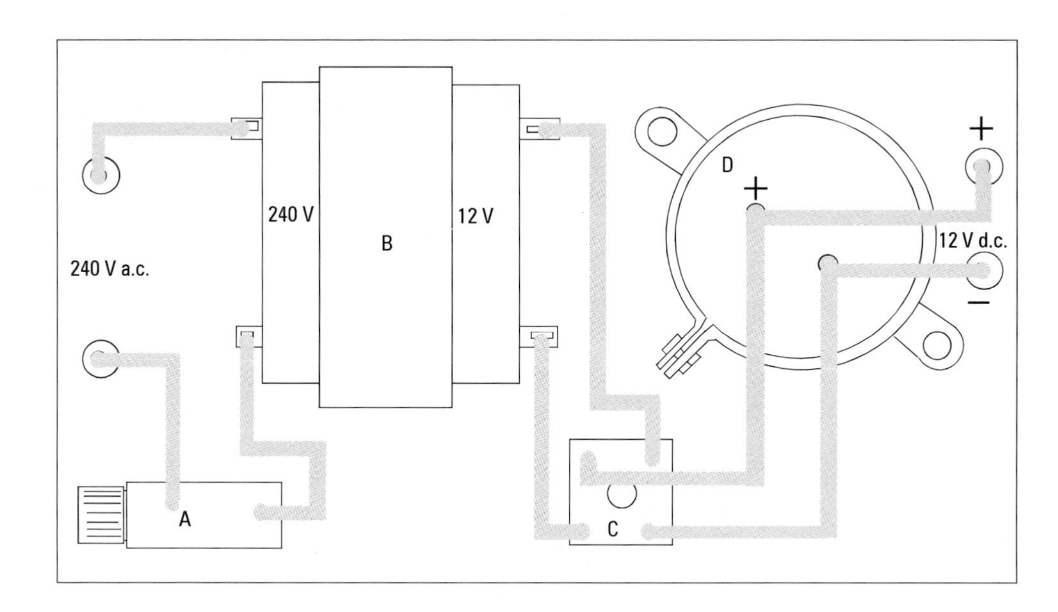


## Parts list

A fuse unit

B 240 V to 12 V double wound transformer

C full-wave rectifier bridge

D smoothing electrolytic capacitor

*Try this*

## Regulated 15-0-15 volt d.c. supply unit

Using the circuit diagram of a regulated 15-0-15 volt d.c. supply draw the following:

(i) the block diagram

(ii) the circuit tracks on the incomplete layout diagram

## Block diagram

## Circuit diagram

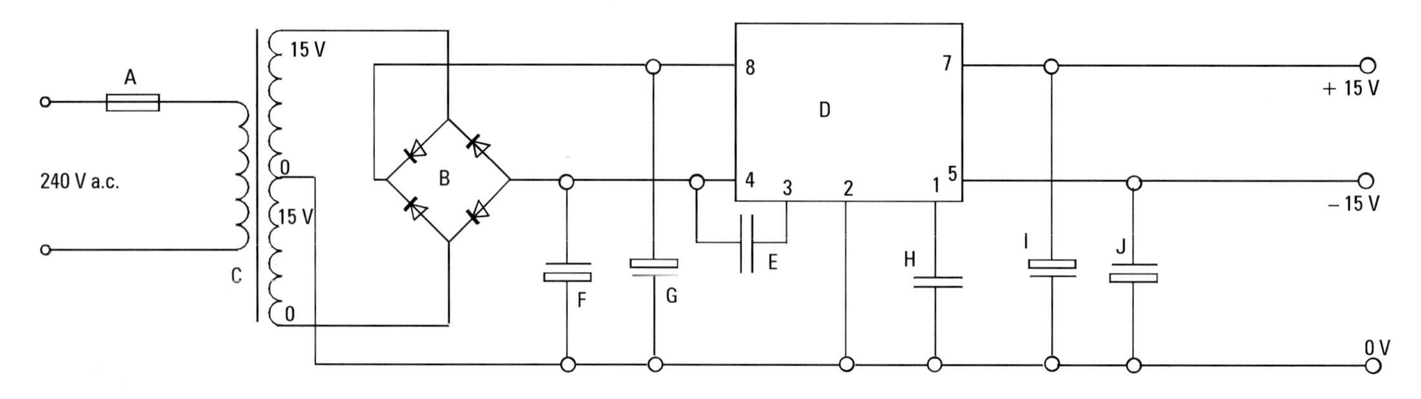


## Layout diagram

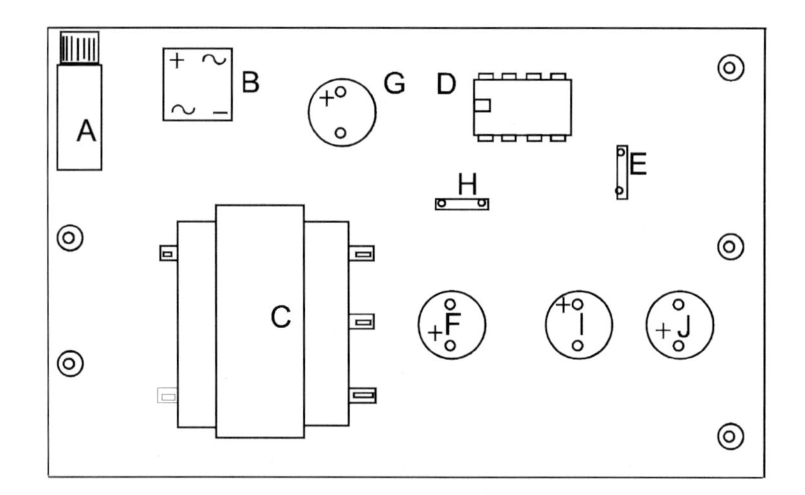


## Parts list

A fuse unit
B bridge
C transformer
D 8 pin DIL regulator RC4195N
E 100 nF capacitor
F 1000 μF 25 V capacitor
G 100 μF 25 V capacitor
H 100 nF capacitor
I 10 μF 25 V capacitor
J 10 μF 25 V capacitor

## Logic diagrams

Logic diagrams use symbols for logic gates. Logic gates are capable of processing digital input signals in order to give an output signal when certain input conditions are met.

Digital signals are a series of pulses which indicate either one of two states, i.e. ON or OFF.

There are three basic types of gate:

- AND
- OR
- NOT

If we consider a two-input gate then an **AND** gate (Figure 6.6) will give an output (ON) only when an input is present at both inputs. (A and B).

A truth table shows the output for all the possible inputs.

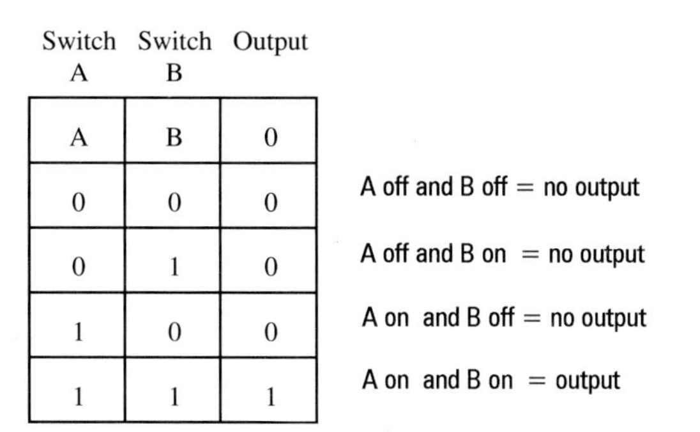

| Switch A | Switch B | Output |
|---|---|---|
| A | B | 0 |
| 0 | 0 | 0 |
| 0 | 1 | 0 |
| 1 | 0 | 0 |
| 1 | 1 | 1 |



An input at both A & B gives an output (ON).

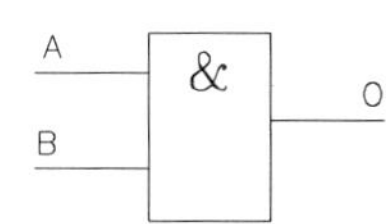


*Figure 6.6* *British Standards Institution symbol for an AND gate*

For an **OR** gate (Figure 6.7) there is an output (ON) when a signal is present at A or B. Truth table for an OR GATE:

| A | B | 0 |
|---|---|---|
| 0 | 0 | 0 |
| 0 | 1 | 1 |
| 1 | 0 | 1 |
| 1 | 1 | 1 |

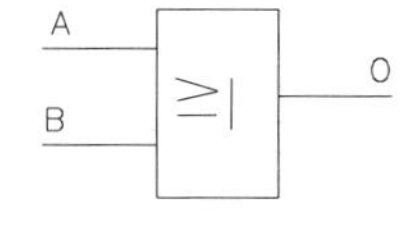


*Figure 6.7* *British Standards Institution symbol for an OR gate*

The **NOT** gate (Figure 6.8) is an inverter which changes the state of the input. In other words it gives an output when there is NOT an input.

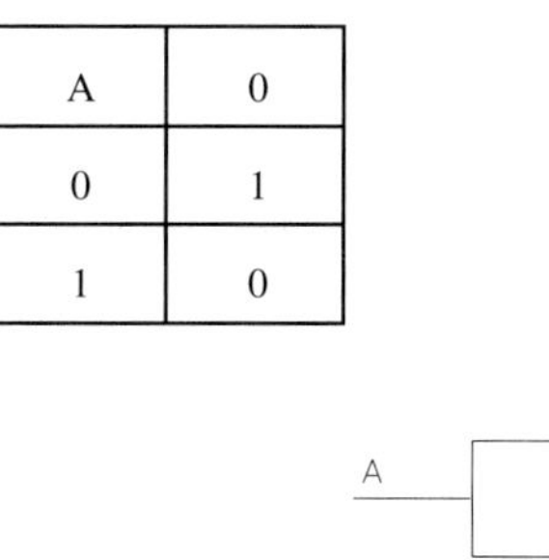

| A | 0 |
|---|---|
| 0 | 1 |
| 1 | 0 |



*Figure 6.8* *British Standards Institution symbol for a NOT gate*

If we put the **NOT** gate with the **AND** gate then we get a **NAND** gate (Figure 6.9) – which is not A and B.

| A | B | 0 |
|---|---|---|
| 0 | 0 | 1 |
| 0 | 1 | 1 |
| 1 | 0 | 1 |
| 1 | 1 | 0 |

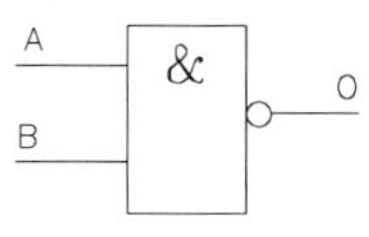


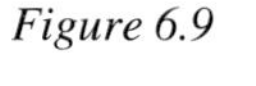

*Figure 6.9* *British Standards Institution symbol for a NAND gate*

If we put the **NOT** gate with the **OR** gate we get a **NOR** gate (Figure 6.10) – which is not A or B.

| A | B | 0 |
|---|---|---|
| 0 | 0 | 1 |
| 0 | 1 | 0 |
| 1 | 0 | 0 |
| 1 | 1 | 0 |

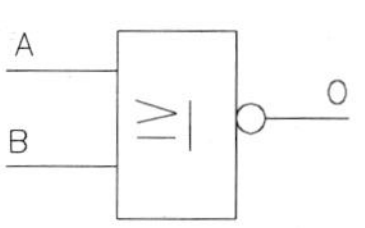


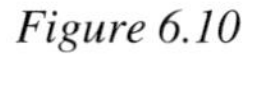

*Figure 6.10* *British Standards Institution symbol for an NOR gate*

Logic gates can be combined to provide a control function which will deliver one or more output signals when certain input conditions are present. This can be shown on a logic diagram (Figure 6.11).

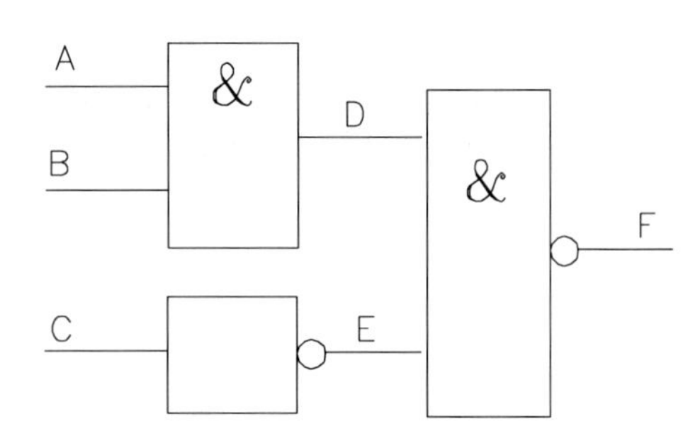


*Figure 6.11* *Three-input logic diagram*

## Points to remember

A block diagram shows a sequence of actions between defined sections.
A circuit diagram allows the operation of a circuit to be understood and uses symbols from BS _______
A layout diagram uses scaled shapes of the actual objects to give a diagrammatic picture of the finished circuit.
Identify the symbols given below:

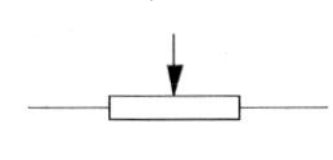

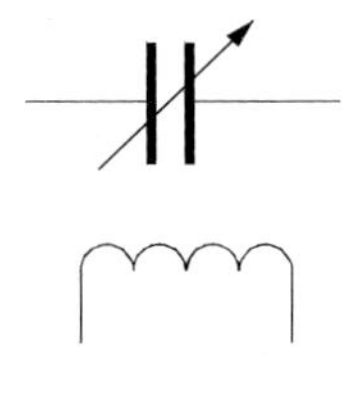

Draw the correct symbols for the following components:

light-dependent resistor

variable capacitor

inductor (general)

zener diode

FET

thyristor

switch

## Self-assessment multi-choice questions

**Circle the correct answers in the grid below.**

1. Which type of diagram uses standard graphical symbols and shows, in detail, which component pins are connected together?
   (a) block
   (b) circuit
   (c) layout
   (d) logic
2. In graphical symbols a variable component would have
   (a) a bar through it
   (b) two arrows pointing towards it
   (c) two arrows pointing away from it
   (d) an arrow through it
3. In graphical symbols a light-sensitive component would have
   (a) a bar through it
   (b) two arrows pointing towards it
   (c) two arrows pointing away from it
   (d) an arrow through it
4. An AND gate will give an output for an input of
   (a) signal present at A and B
   (b) signal present at A, not at B
   (c) signal present at B, not at A
   (d) no signal present at A or B
5. A NOR gate will give an output for an input of
   (a) signal present at A and B
   (b) signal present at A, not at B
   (c) signal present at B, not at A
   (d) no signal present at A or B

### Answer grid

| | | | | |
|---|---|---|---|---|
| 1 | a | b | c | d |
| 2 | a | b | c | d |
| 3 | a | b | c | d |
| 4 | a | b | c | d |
| 5 | a | b | c | d |

# 7

# Electronic Maintenance, Repair and Servicing

Complete the following to remind yourself of some important facts from the previous chapter.

Name the different types of diagram that were considered in the previous chapter.

If it is impossible to include all the relevant details on a diagram, what kind of drawing with a key can help to give the necessary information?

What sort of images are used on a layout diagram?

**On completion of this chapter you should be able to:**

- recognise some of the types of connectors used in the electronics industry
- recognise the important techniques of soldering
- state safety precautions to be adopted when using soldering equipment
- describe methods of connecting and/or replacing wires and components to printed circuit boards
- describe the use and care of tools required for connecting electronic components
- complete the revision exercise at the beginning of the next chapter

## Part 1

**Safety**
**Take care when handling components! Remember that you can get a shock from capacitors that have remained charged. Also that your body can hold a very high static charge of electricity which can destroy electronic components should it become discharged through them.**

### Tools and their uses

Shown in the photograph below are some of the general tools you will come across when working with electronic components. Remember to keep all tools in a good condition ready for use. Always choose the correct tool for the job. For instance, the use of installation pliers to bend and insert a 0.4 W resistor into a PCB is obviously wrong!

An electrician's tool kit is likely to contain basic items such as lightweight pliers, cutters and screwdrivers suitable for use with electronic equipment. Some examples are shown in Figure 7.1.

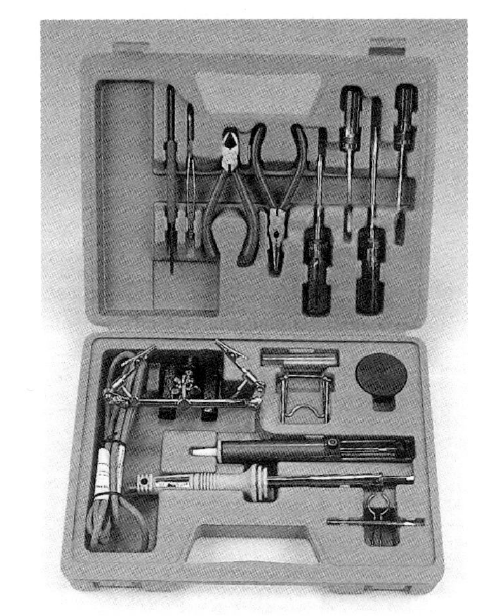

*Figure 7.1*

# Connections

Connections fall into two categories

- readily separable by unplugging (connectors)
- mechanically fixed, such as soldered and wire-wrapped

# Connectors

## Plugs and sockets

There are many types of plugs and sockets that are used in electronics to connect sections together. These range from simple two-pin arrangements to complex multipin and screened connections.

A few examples of plugs and sockets are given below. A jack socket and plug (Figure 7.2) have many applications in audio equipment. The cable connections may be terminal screw or solder.

So as to ensure good electrical connections the contacts are often gold-plated silver (Figure 7.3). The cable connections to this type of plug are made on to solder terminals.

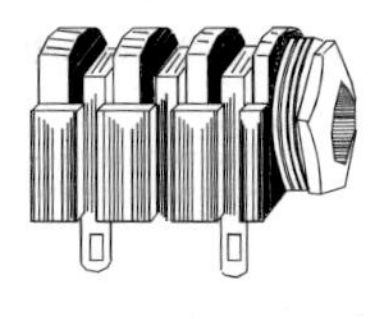

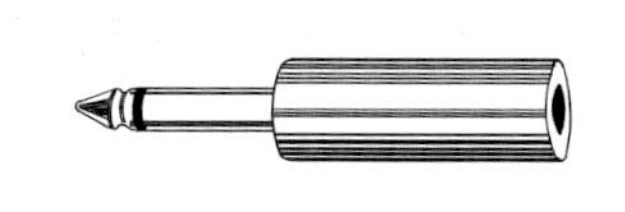

*Figure 7.2* *Jack socket and plug*

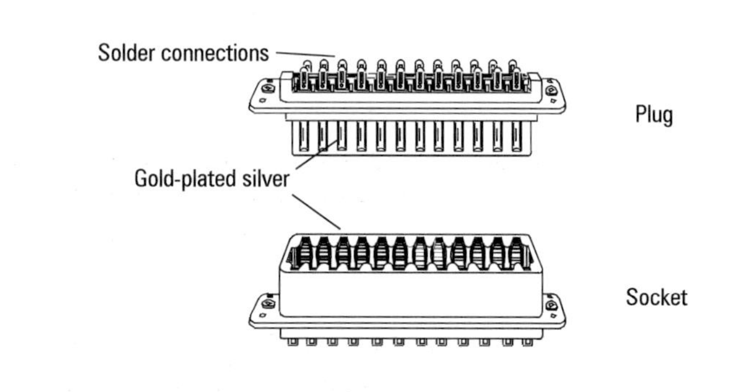


*Figure 7.3* *24 way module connector*

Coaxial sockets and plugs (Figure 7.4) are used on screened cables where the screening is connected to the outside metal and the core to the centre pin.

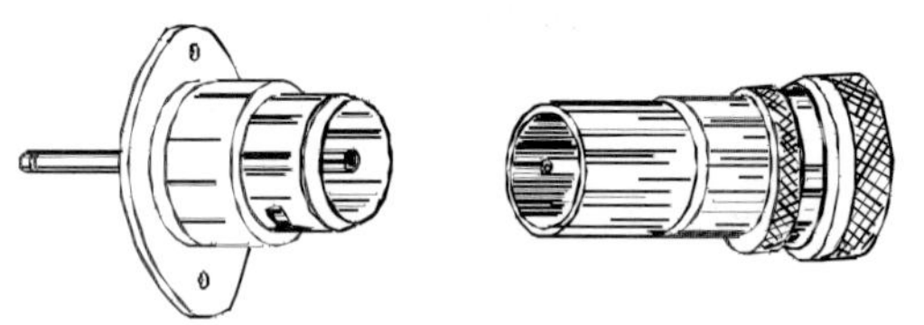

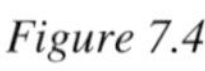

*Figure 7.4* *Coaxial socket and plug*

Some PLCC sockets and/or d.i.l. devices will require a special tool (Figures 7.5 and 7.6) when extracting the component from the socket. When removing a component remember to note its orientation (which way round it is) so that a replacement can be readily fitted.

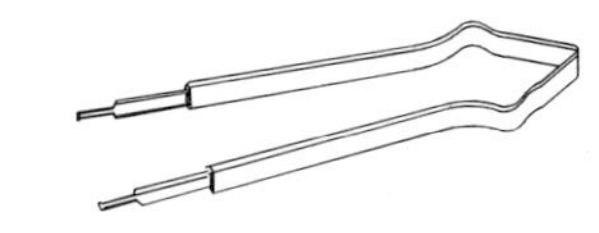

*Figure 7.5* *Tool for extracting d.i.l. component from socket*

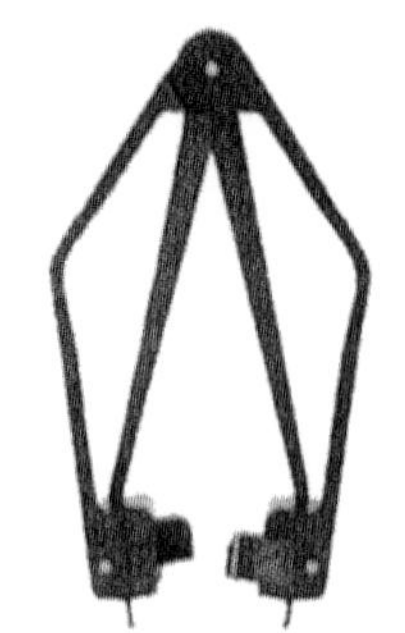

*Figure 7.6* *Tool for extracting PLCC component from socket*

When removing multiple connectors that may be interchangeable, always note where they were removed from so that they can be replaced correctly.

## Connecting to printed circuit boards

Where printed circuit boards are made as a replaceable unit they often incorporate a plug-in arrangement on the board. One example of this is the edge conncctor (Figure 7.7). Here the printed circuit board has been made so that the track goes out to the edge of the board. To ensure a good electrical contact the area that plugs into the socket is often gold-plated.

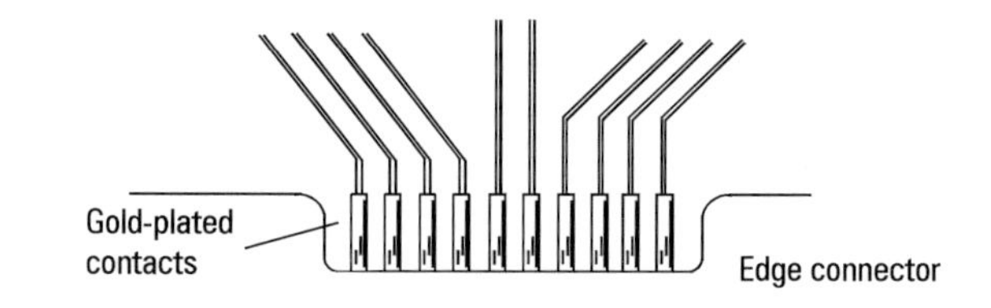


*Figure 7.7* *Edge connector on PCB*

Printed circuit boards should always be handled with care. The manufacturing process of the board means that they may be very thin in places and can crack easily. Consideration must also be given to the fact that many of the components on the boards may be damaged by the voltages in the human body. The boards must be handled by their edges without touching the conductor tracks. It may be necessary to wear an earthed strap to stop the effects of static electricity from damaging the sensitive components.

Circuit components or their mounting bases are connected directly on the board by soldering. Great care must also be exercised when replacing the components which have been mounted on a PCB. The copper tracks are very fine and can be fractured if the board is bent or flexed during the process.

**If voltages above 50 V a.c. and 120 V d.c. are in use the supply lines should be connected to the socket NOT the plug.**

*Try this*

Using a manufacturer's catalogue, find the section on connectors and list those that have not been mentioned and give an example for their use.

| Connector | Use |
|---|---|
| | |

# Mechanically fixed components

## Connector blocks

Connector blocks provide a means of connection and disconnection of cables from a PCB. The blocks can be one, two, three or five-way versions and the wires are secured by screws, clamps or soldering depending on the type and the reason for the connection. Wires attached by screws facilitate easy disconnection when servicing, testing or replacing, while soldered connections are suitable for more permanent installations. Most types of connector block feature wire guards under each screw to protect the wire from any twisting or cutting action by the screw.

*Figure 7.8*

## Wire wrap connections

As its name implies this is a method that involves wrapping a wire round a pin to form the connection. The pin is inserted into the printed circuit board, as shown in Figure 7.9. This is usually soldered to the copper track to form the electrical connection

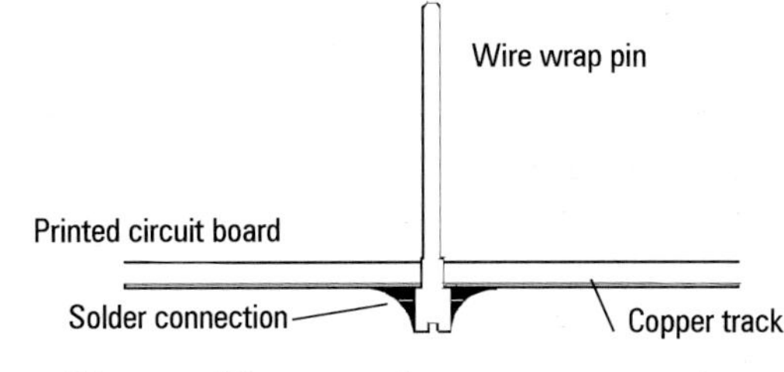

*Figure 7.9 Wire wrap pin*

The pin is rectangular in shape with fairly sharp edges (Figure 7.10). The wire is wrapped tightly round the pin by means of a special tool.

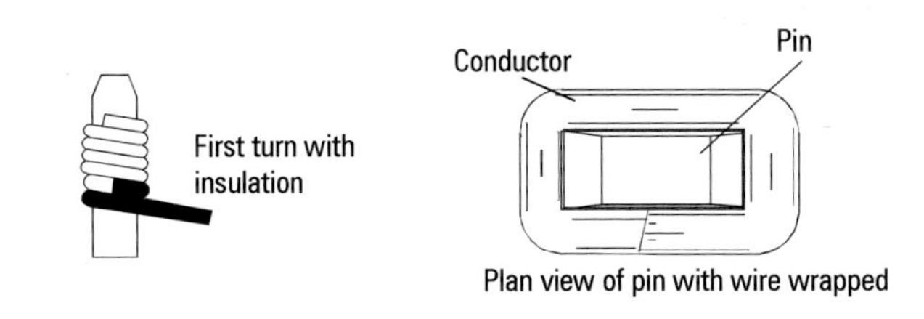


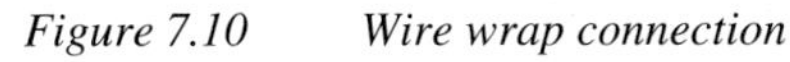

*Figure 7.10 Wire wrap connection*

There are many tools available to perform this connection. These may be hand-operated or electrically powered. The wire is usually fed through the centre of the tool and into the wrapping bit.

A typical manually operated tool (Figure 7.11) would be loaded with the wire and then placed over the terminal post. The tool would then be rotated about eight turns. Usually this produces a joint with the first turn of insulation complete, as shown in Figure 7.10. This greatly increases the ability of the joint to withstand vibration.

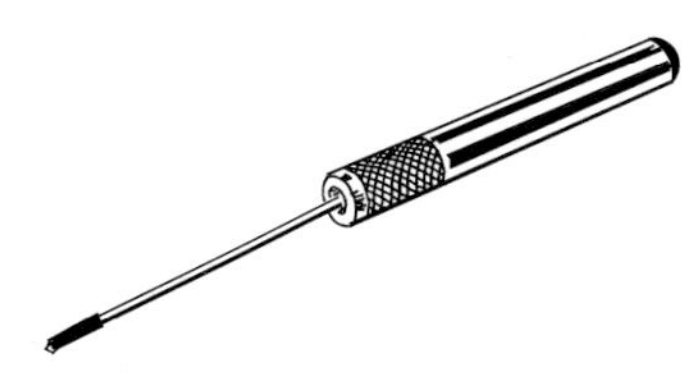

*Figure 7.11 Wrapping tool*

A power-operated tool carries out the rotation automatically. Some will also strip and cut off excess wire.

The first termination on a pin is kept to the bottom so that other connections may be made above if necessary (Figure 7.12).

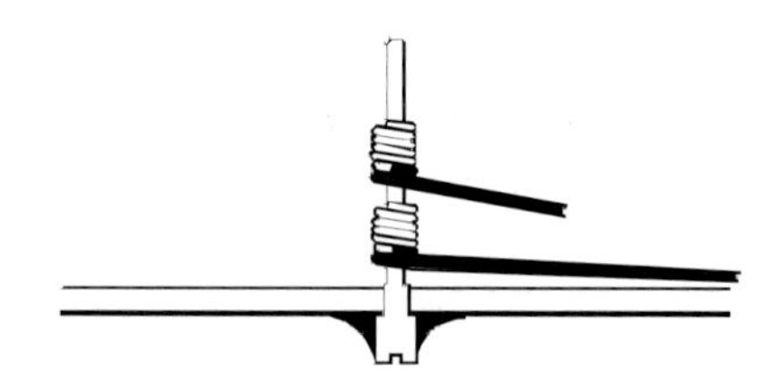

*Figure 7.12 Wire wrap pin with connections*

There is also a tool available for unwrapping the joint.

## Insulation displacement connection (IDC)

There are methods of connection which do not require the removal of the insulation from a cable.

The connection is made by pressing the cable onto a V-shaped blade (Figures 7.13 and 7.14). This blade pushes through the insulation and then touches the conductor. The blade is shaped so as not to damage the conductor.

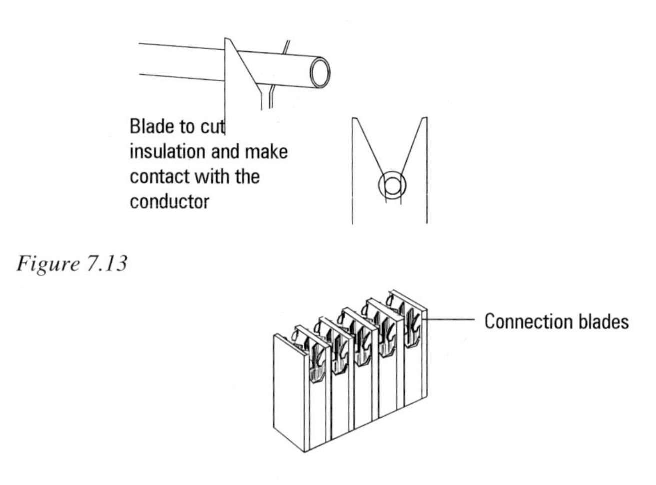


*Figure 7.13*

*Figure 7.14 Socket with insulation displacement connection*

These are used extensively for ribbon cables (Figure 7.15) where all of the conductors can be connected at the same time with the minimum separation of individual cores.

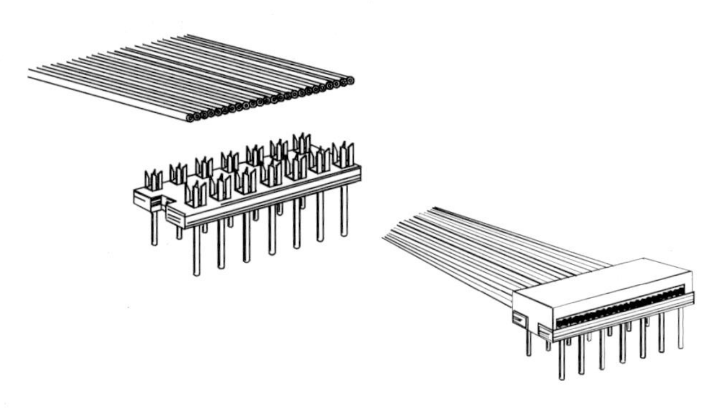

*Figure 7.15 Ribbon cable connection*

To ensure good electrical and mechanical connections special tools have been developed for inserting the cables into the connectors. Figure 7.16 shows a typical hand-held single core tool. Where there are a lot of connections to be made, more sophisticated multicore press tools are available.

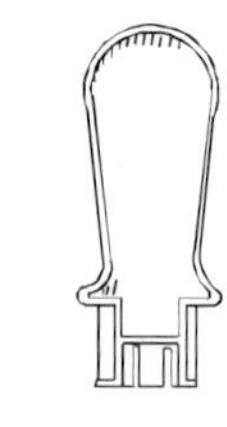

*Figure 7.16 Single core tool*

## Crimp

A crimp tool is available for crimp connections. It connects an individual wire to a contact which connects to a connector housing (Figures 7.17 and 7.18). There is also a tool available for removing the contact from the connector body. Some examples are shown. It is likely that you will find that manufacturers have their own specific tools for their own products.

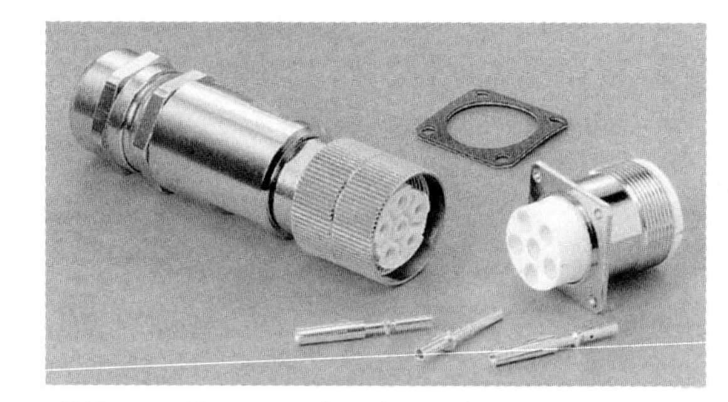

*Figure 7.17 Connector housing and contacts*

*Figure 7.18 Crimp and extraction tools*

### *Points to remember*

Connections fall into two categories

- readily separable by ______________________
- mechanically fixed, such as ______________________

Circuit components are connected directly to a printed circuit board by ______________

To ensure a good electrical contact the plug-in edge connector is often ________ plated.

# Part 2

## Solder connections

The most-used connection is the soldered one. This generally uses an alloy of tin and lead to bond metallic materials together. The alloy used for electrical soldering is 60% tin and 40% lead, and this melts at 188 °C. To ensure a good electrical and mechanical connection a resin flux is used to prevent oxidation of the metals. The solder in general use has the flux in cores that run through the length of the solder (Figure 7.19). There are other solders that exist for a variety of temperatures and purposes. For example, in response to the needs of a "greener" environment there is now a lead-free solder and a flux that is halide free.

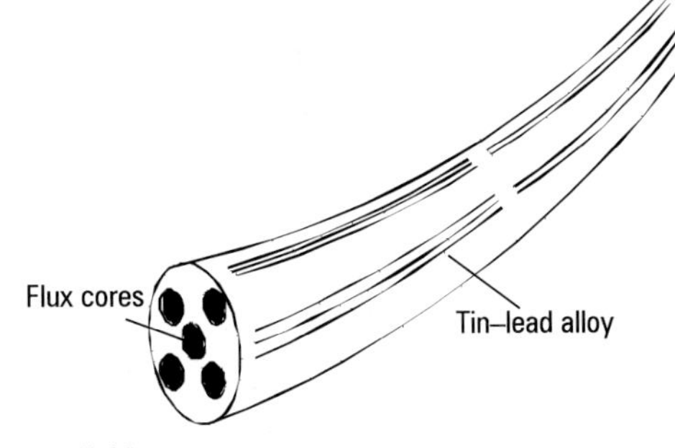


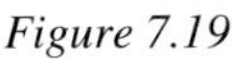
*Figure 7.19*

## Soldering irons

The type and size of soldering iron (Figure 7.20) that should be used must be related to the connections that have to be made.

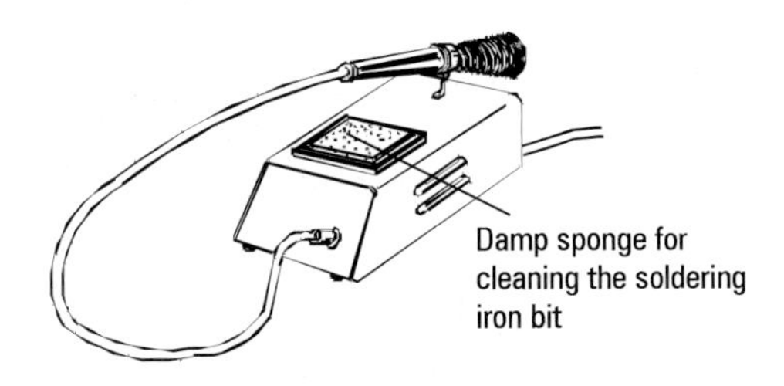


*Figure 7.20*

Some soldering irons operate at a fixed temperature and others can be variable. There are also different sizes and shapes of bit, which are interchangeable. For solder to flow on all of the surfaces that are being connected, all the metal must reach a temperature of at least 190 °C. If there are large metal contacts that have to be heated then an iron capable of raising and maintaining the temperature must be used. It is very easy to apply too much heat, especially where there are plastics in use. Heat transmitted through the metal will soon melt insulation or terminal bodies. An iron of suitable dimensions and temperature must be used.

> ***Remember***
> **The type and size of soldering iron that should be used must be related to the connections that have to be made.**

## Soldering techniques

There are two stages to making a good solder connection. First, the separate parts of the joint must be prepared, and secondly they need to be put together and soldered.

## Preparation

The key to a well-prepared connection is care and neatness. Time taken here can save time later.

There are two types of lead wires used:

- solid single strand
- multi-stranded

Solid single-strand wire should have any insulation removed to the required length, and then the conductor is usually ready for connection.

When using a stranded cable, in addition to removing the insulation, the copper strands need to be twisted together and "tinned". To tin the wire the soldering iron and solder are applied to the end of the twisted wire and then worked up towards the insulation (Figure 7.21).

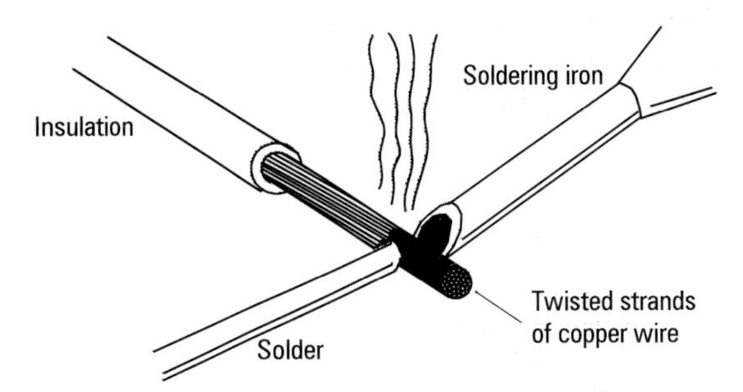


*Figure 7.21*

This is a skill that needs practice, for the amount of solder and heat used is critical. The result of a good tinned wire should be a bright shiny surface where all strands are joined together as one.

Part of the preparation of any joint is forming the lead wires to the required shape. There are several methods of soldering wires to solid terminations; three of these are now dealt with in some detail.

## Soldering to a pin

When connecting to a pin, the wire should be bent to a hook shape using round-nosed pliers, as shown in Figure 7.22. The hook should be placed over the pin and then squeezed to form a tight fit.

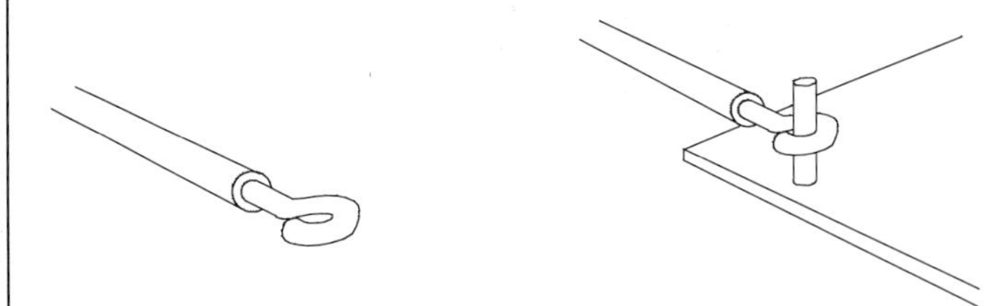

*Figure 7.22 Soldering to pin*

## Soldering to tag strip

The connecting wire is threaded through the hole in the tag strip and then bent round the tag, as shown in Figure 7.23. Excess wire is then cut off and the joint squeezed flat with pliers.

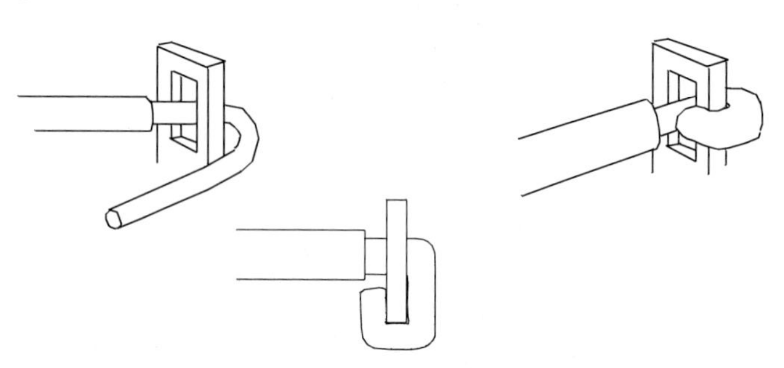

*Figure 7.23 Soldering to tag strip*

## Soldering to printed circuit board

If it is a component to be fitted to the printed circuit board, the leads must be formed to the correct pitch before being inserted (Figure 7.24). This can be carried out using a pair of round-nosed pliers or special lead-forming tools. The lead-forming tool consists of two arms over which the component leads are bent. When a number of components all have to be prepared the lead-forming tools ensure a consistent lead pitch on all components.

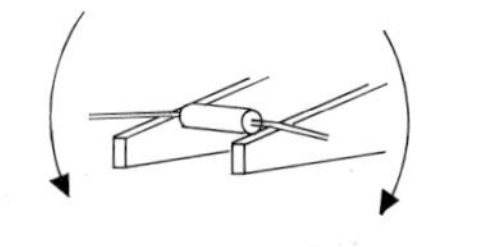

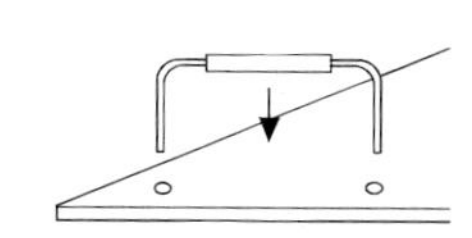

*Figure 7.24*

After the component has been inserted into the board the leads should be bent to keep the component in place. The leads can now be cut back, leaving enough to hold the component but not too much so as to short out other tracks on the printed circuit board (Figure 7.25).

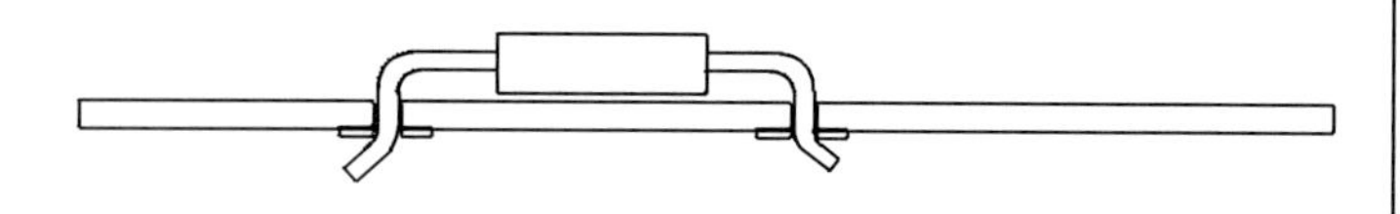

*Figure 7.25*

There are three main factors for a good connection.

- cleanliness
- correct heat
- correct flux (if a cored solder is used this should be ensured)

Remember that cleanliness at the joint can be increased by tinning the surfaces first; that is, by coating them in a thin layer of solder.

Aluminium conductors cannot be soldered in this way.

## For good soldering joints

- the iron should be clean and fluxed
- when solder is applied to the iron it should appear bright and shiny if the iron is up to temperature
- the surfaces to be soldered should be clean and tinned
- the iron should be applied to the surfaces to be connected (Figure 7.26), *not* to the solder
- solder should be seen to run over the surfaces when the joint is made
- the joint should be allowed to cool, *not* cooled by blowing or applying a damp cloth

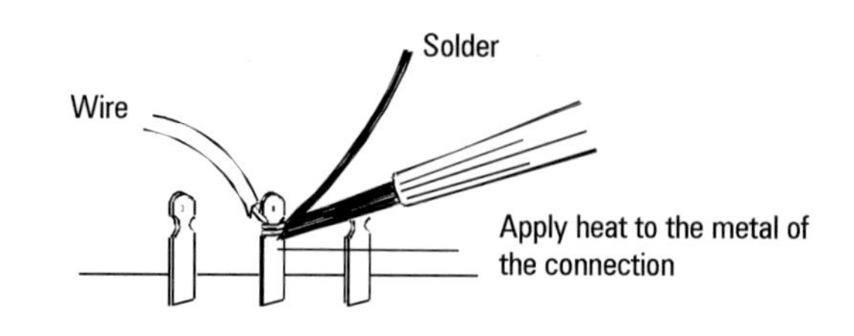


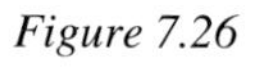

*Figure 7.26*

## Precautions when soldering

- never flick off solder from the iron as this may cause harm to persons, clothing or apparatus
- where soldering is done inside equipment, it may be necessary to cover components
- heat sinks (Figure 7.27) may have to be applied on some heat-sensitive components
- always keep the soldering bit clean and tinned
- use the soldering iron stand to prevent danger from burns or fire.

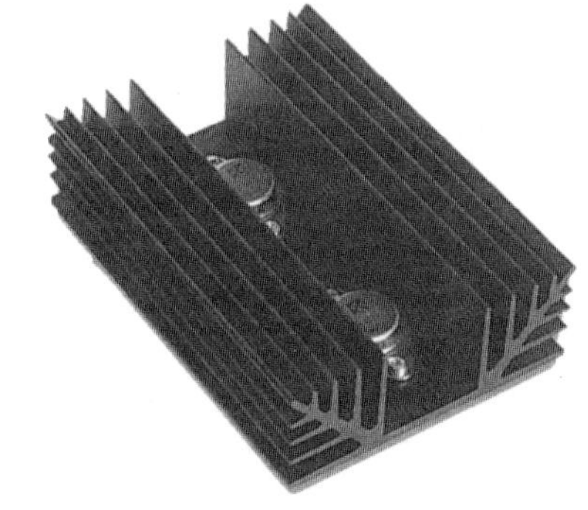

*Figure 7.27 Heat sink*

## Common soldering faults

- **Dry joints** – These are caused generally by not using a high enough temperature or by moving the leads before the solder has "set". All surfaces have to get to a temperature high enough to melt the solder. A dry joint often looks grey and dull, not bright and shiny as it should be.
- **High-resistance joints** – dry joints are often also high-resistance joints. It is not always possible to detect a high-resistance joint visually. Where the surfaces have not been cleaned and tinned correctly, it is possible for a crystallisation to form between the surfaces, particularly if a solder was used without sufficient flux.

- **Excess solder** – This can lead to connections being shorted out accidentally.
- **Too much heat** – This can cause several problems:
  - damage to insulation
  - damage to the printed circuit board
  - hardening of the wire making it brittle
  - damage to components: components can be protected by using a shield or a heat shunt (Figure 7.28).

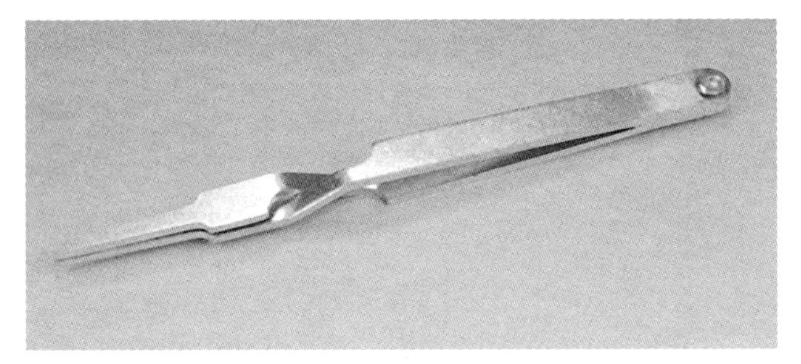

*Figure 7.28* *Heat shunt – when clipped to a transistor lead will prevent excess heat reaching the device*

## Desoldering techniques

Often when making or repairing electronic equipment it is necessary to remove a component from a printed circuit board. In order to avoid damage to the rest of the board, cut away the component to be discarded before de-soldering. The technique of desoldering is used to remove the solder from the joint. Two methods are commonly used. The simplest to use is the desoldering braid. The braid is placed on the joint, then the soldering iron is placed on top to heat both the braid and the joint. When hot, the solder is wicked up into the braid (Figure 7.29).

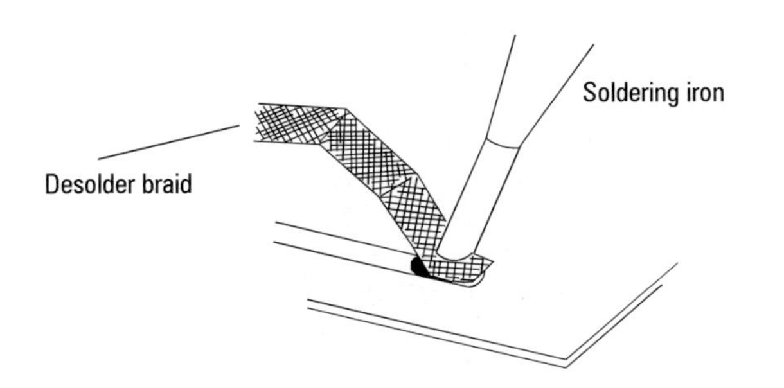


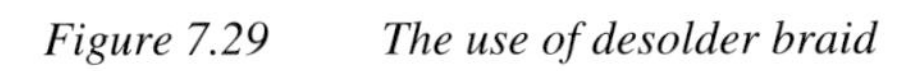

*Figure 7.29* *The use of desolder braid*

The second method is to use a desoldering tool. This has a spring-loaded plunger that sucks hot solder from the joint when the plunger is released (Figure 7.30). The holes in the boards should be cleared and everything completely clean before mounting the new component.

If excessive heat is applied when de-soldering, the tracks can become detached from the board and broken. In addition to this, the components themselves can be damaged by heat.

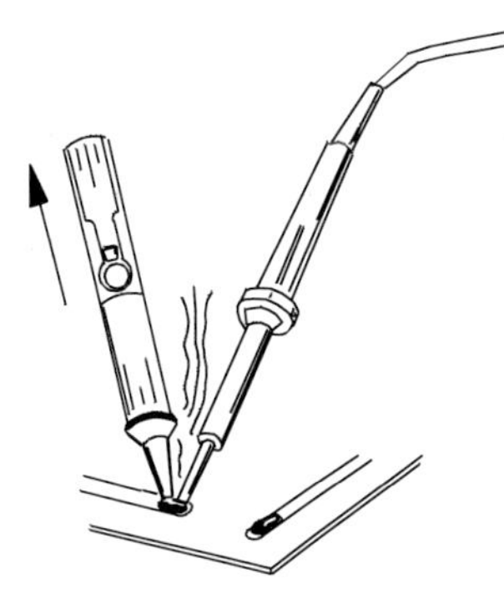

*Figure 7.30* *The use of a desolder tool*

## Common desoldering problems

- When the desoldering tool is full of solder it does not operate smoothly.
- When the joint is not hot enough, solder remains on the joint. If this happens it is often better to re-solder the joint before desoldering again.
- When the joint is too hot the PCB tracks may lift from the board or the component be damaged.
- When the desolder nozzle does not seal to the board little suction is achieved.
- Some people leave the iron on the joint while applying the desolder nozzle – this can work, with care, but tends to damage the nozzle and the PCB as the desolder nozzle hammers the iron into the PCB surface.

## Surface mount devices

Repairs to the soldering of SMD should only be undertaken using the appropriate tools and by those who have received the appropriate training.

## *Points to remember*

What are the two stages in making a good solder connection?

When soldering a stranded cable, remove the insulation and _________ the copper strands together. To tin the wire, apply the soldering iron and solder to the ends of the twisted wire and work up towards the insulation.

When soldering to a PCB use a ____________ tool to form the correct pitch on the leads.

## *Try this*

Name the connections or tools shown below.

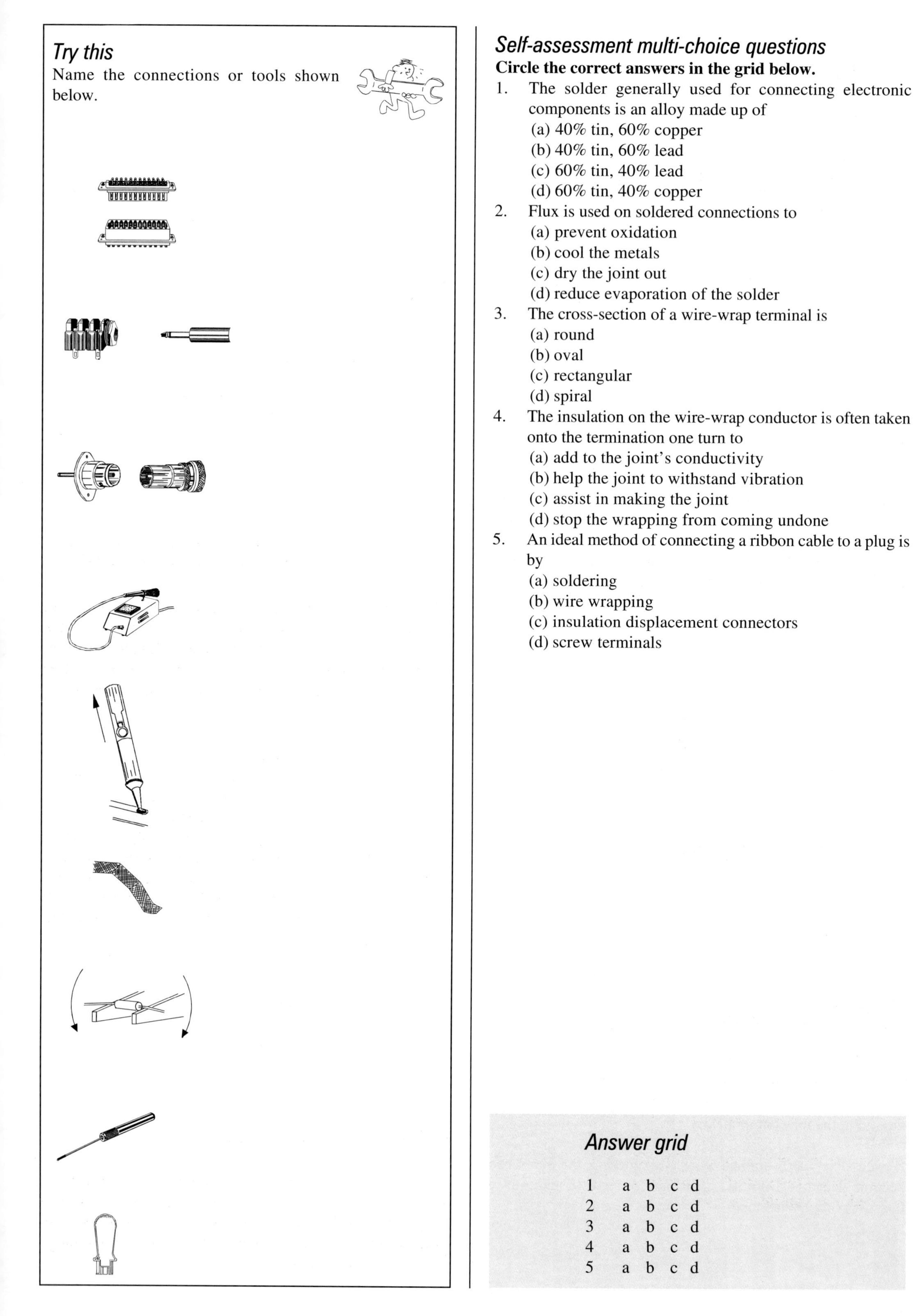

## *Self-assessment multi-choice questions*

**Circle the correct answers in the grid below.**

1. The solder generally used for connecting electronic components is an alloy made up of
   (a) 40% tin, 60% copper
   (b) 40% tin, 60% lead
   (c) 60% tin, 40% lead
   (d) 60% tin, 40% copper
2. Flux is used on soldered connections to
   (a) prevent oxidation
   (b) cool the metals
   (c) dry the joint out
   (d) reduce evaporation of the solder
3. The cross-section of a wire-wrap terminal is
   (a) round
   (b) oval
   (c) rectangular
   (d) spiral
4. The insulation on the wire-wrap conductor is often taken onto the termination one turn to
   (a) add to the joint's conductivity
   (b) help the joint to withstand vibration
   (c) assist in making the joint
   (d) stop the wrapping from coming undone
5. An ideal method of connecting a ribbon cable to a plug is by
   (a) soldering
   (b) wire wrapping
   (c) insulation displacement connectors
   (d) screw terminals

### *Answer grid*

| | | | | |
|---|---|---|---|---|
| 1 | a | b | c | d |
| 2 | a | b | c | d |
| 3 | a | b | c | d |
| 4 | a | b | c | d |
| 5 | a | b | c | d |

# 8

# Measurements

Complete the following to remind yourself of some important facts from the previous chapter.

Electronic component connections include those that are readily separable, such as:

and those that are mechanically fixed such as wire wrapped or

To ensure a good electrical contact the area of a PCB that plugs into the socket is plated with what material?

Explain, using a diagram if it helps, how to solder to a resistor into a PCB. List any tools that you would require.

The alloy that is used for electrical soldering melts at 188 °C. What are the constituent parts and in what proportion?

List four common soldering faults.

**On completion of this chapter you should be able to:**

- ◆ describe how to carry out tests on resistors and diodes
- ◆ measure the voltages across resistors in a circuit
- ◆ measure the current flowing in a circuit
- ◆ relate the r.m.s. value of a sine wave to that of the peak value
- ◆ compare sine and square waveforms
- ◆ describe the use of the oscilloscope (CRO)
- ◆ identify waveform characteristics

# Part 1

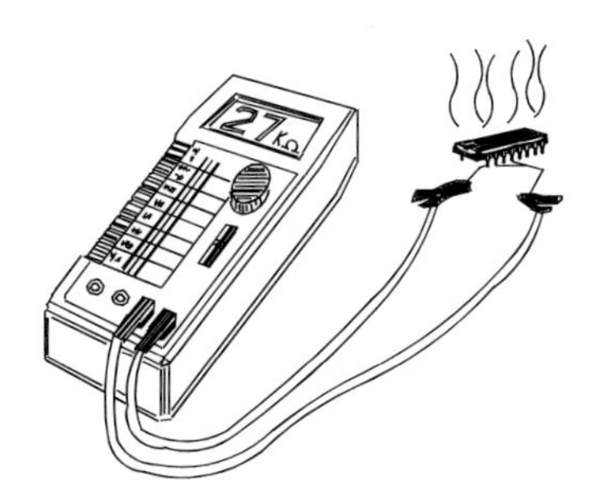


*Figure 8.1* *Use the correct instrument on a suitable scale and voltage for the component to be tested!*

When using test instruments (Figure 8.1), take care not to use voltages and currents in excess of the component's rated value, or the component may be destroyed. Before using the measuring instrument, check that it has the standard certification. This confirms that the calibration is correct.

## Measuring resistance

The instrument used to measure resistance is an ohmmeter (Figure 8.2). As the values of resistance can range from a few ohms up to millions of ohms, the ohmmeter must also have ranges capable of measuring this. Traditionally, multi-range instruments have been used, where the ranges are switched as required. More recently, self-ranging digital instruments have come into use. Whichever type of instrument is used it is important to have some idea as to what the reading should be so that you are only checking values.

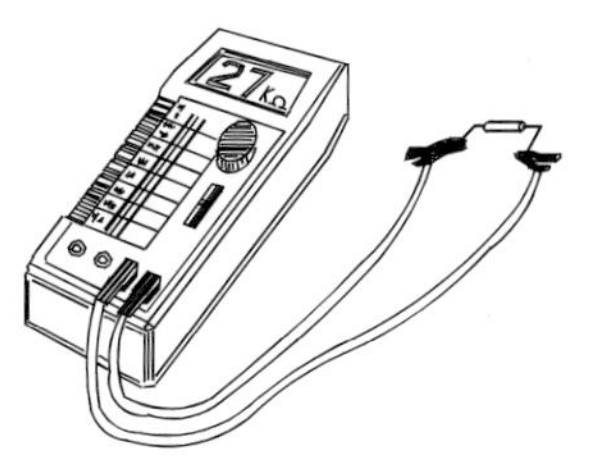


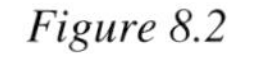

*Figure 8.2* *Measuring resistance*

Before measuring resistance it is important to first check the condition of the internal meter battery. If this is low it can lead to inaccurate readings. Secondly, check that the meter has been set to zero with the leads connected together and the range switched to that required.

## Analogue readings

Measuring resistance with a meter which has an analogue scale can sometimes be confusing. Figure 8.3 shows a typical resistance scale on a multimeter. The scale is not linear, and the zero is at the opposite end of the meter to the other scales.

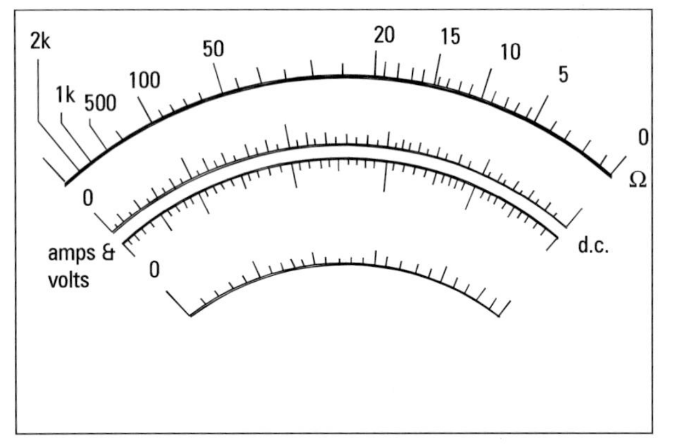

*Figure 8.3*

## *Try this*

## Measuring resistance

Write the reading of each meter in Figures 8.4 and 8.5 in the space provided. The range switch setting shows the maximum that can be indicated, and the reading must be adapted accordingly.

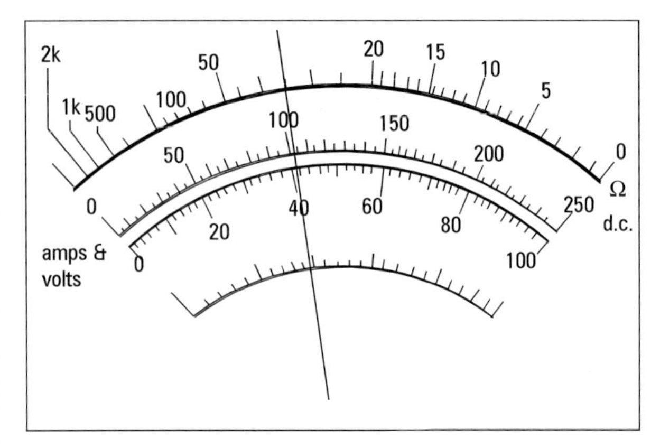

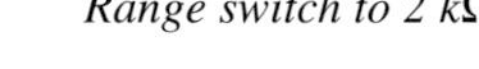
*Figure 8.4* *Range switch to 2 kΩ*

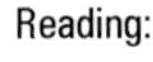
Reading: ________

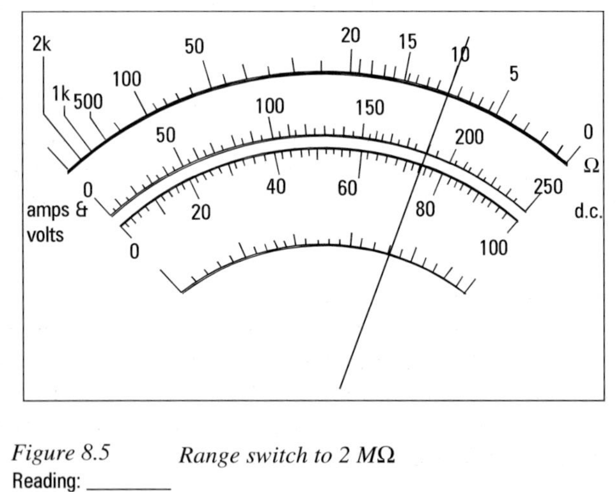


*Figure 8.5* *Range switch to 2 MΩ*

Reading: ________

## Testing diodes

A diode is a device which has a low resistance when a voltage is applied in the forward direction (Figure 8.6) and a high resistance when applied in the reverse direction (Figure 8.7).

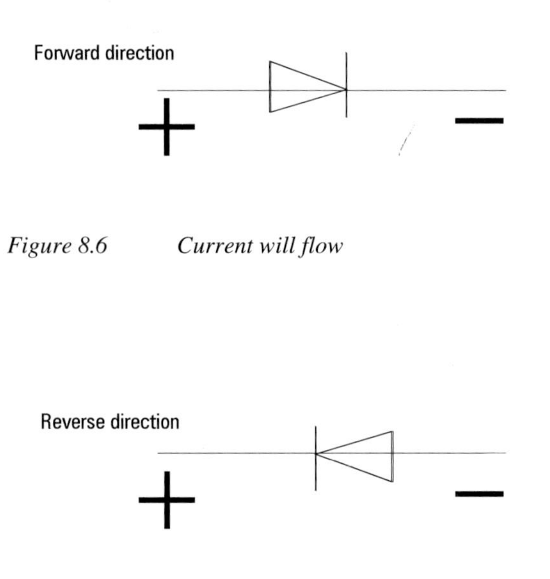


*Figure 8.6* *Current will flow*



*Figure 8.7* *Current will not flow*

Using the same instrument as used for measuring resistance it is possible to check if a diode is working correctly.

The ohmmeter has its own internal battery and this supplies the voltage to apply across the diode. The ohmmeter has its own polarity and this must be known to correctly apply the forward or reverse voltage.

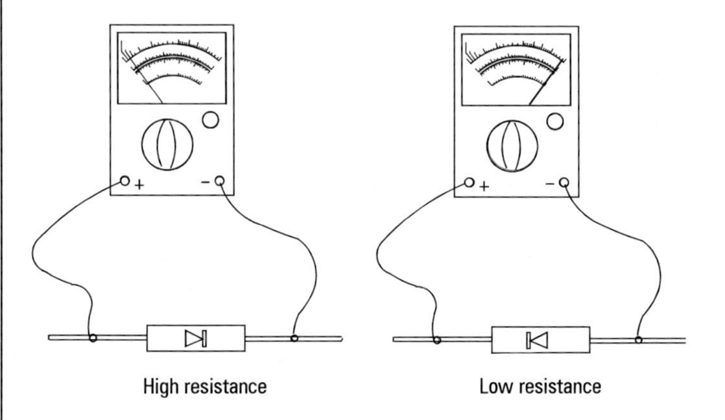


*Figure 8.8*

On many analogue meters using the resistance range the negative terminal is more positive than the positive terminal. This means that the meter is connected to the diode the opposite way than it would at first appear (Figure 8.8).

Digital meters have a very low voltage output, and unless they have a diode testing range they are not always suitable for this purpose.

## *Try this*

## Testing diodes

Write the reading of each meter in Figures 8.9 and 8.10 in the spaces provided and state whether the diode is working satisfactorily.

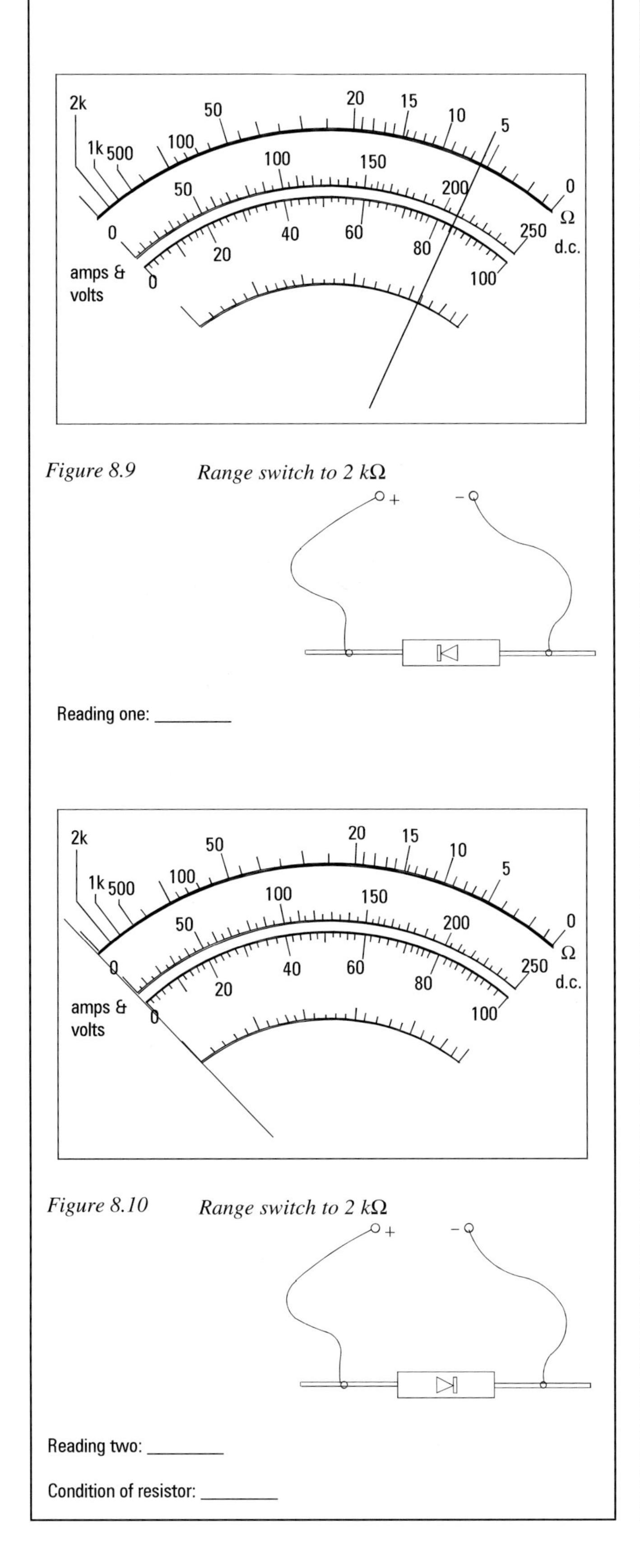


*Figure 8.9* *Range switch to 2 kΩ*

Reading one: ________

*Figure 8.10* *Range switch to 2 kΩ*

Reading two: ________

Condition of resistor: ________

## Testing transistors

As transistors consist of pnp or npn configurations (Figure 8.11), the testing of them is similar to diodes.

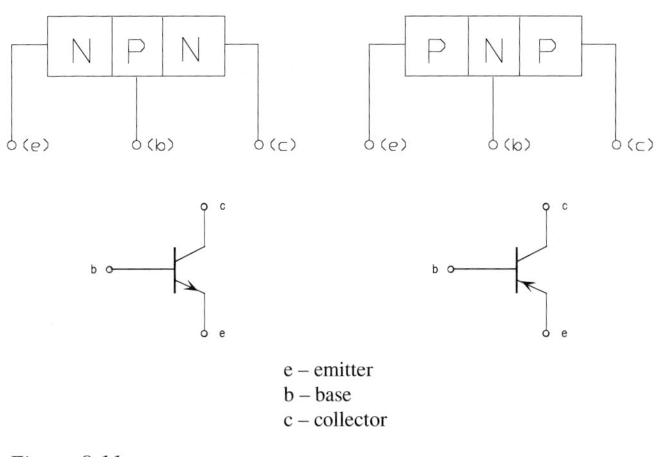


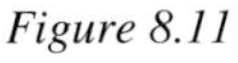

*Figure 8.11*

There are special meters with three terminals for checking transistors, but an ohmmeter can be used for determining whether a transistor is conducting correctly.

When connecting an ohmmeter to an npn transistor, as shown in Figure 8.12, the meter should indicate a high resistance. Reversing the meter connections to "b" and "c" should produce a low resistance. Similar results should be obtained if the meter is connected between "c" and "e".

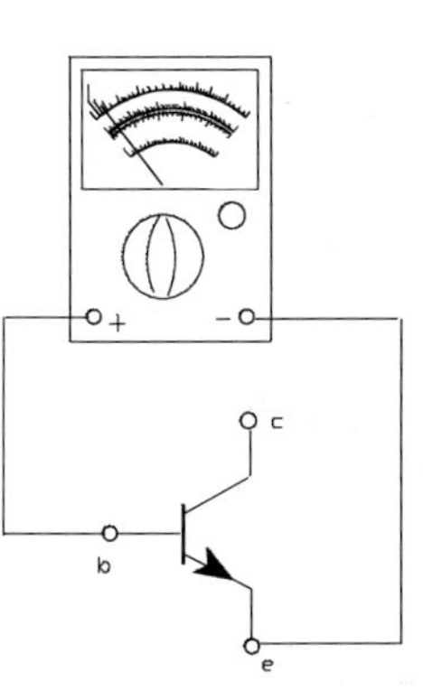


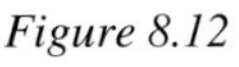

*Figure 8.12*

Testing a pnp transistor is similar, but the low-resistance readings will be obtained when the polarity is the reverse of that of an npn.

The simple go/no-go transistor test may not be sufficient for proper diagnostic testing and a commercial tester may have to be used. This will mean that the transistor will have to be removed from the circuit, but the test result will be far more comprehensive.

More complex devices, such as integrated circuits, are normally tested by replacement with another similar device. This should only be carried out if the circuit conditions are correct and in accordance with the manufacturer's specification, otherwise the replacement device may well end up in the same damaged state as the original.

## Try this

### Testing transistors

Write the reading of each meter in Figures 8.13 and 8.14 in the spaces provided and state whether the transistor is working satisfactorily.

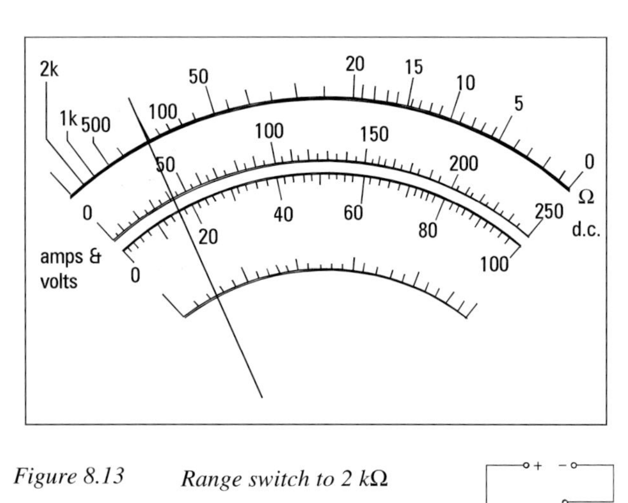


*Figure 8.13* *Range switch to 2 kΩ*

Reading one: ________

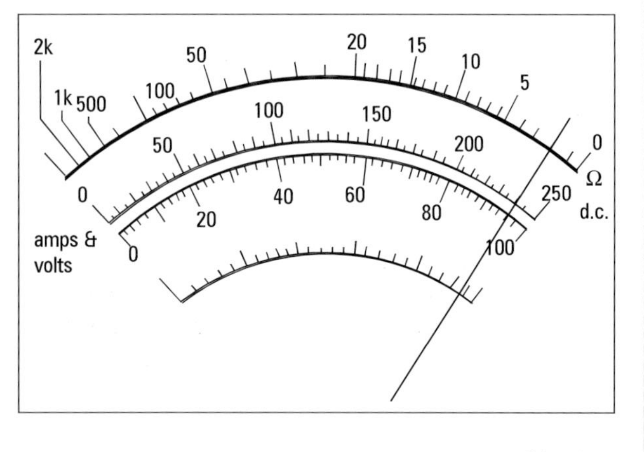


*Figure 8.14* *Range switch to 2 kΩ*

Reading two: ________

Condition of resistor: ________

*Points to remember*

### Electronic measurements

The instrument used to measure resistance is a

Before measuring resistance check the following:

- the condition of the internal battery
- that the meter has been set to ________ with the leads connected together
- that the correct range has been selected

A diode will have a _________ resistance when a voltage is applied in a forward direction.

A diode will have a _________ resistance when a voltage is applied in a reverse direction.

The ohmmeter has its own polarity and this must be known in order to correctly apply the forward or reverse voltage.

The testing of transistors is similar to the testing of ________

Transistors can be checked with special meters with three terminals or with a ________________________

How are integrated circuits normally tested?

# Part 2

## Measuring voltage

The instrument used for measuring voltage may be the same one as used for resistance, but switched to the voltage range. It may, however, be a completely independent voltmeter.

Whenever it is necessary to take voltage readings care must be taken, as the readings have to be carried out when the circuit is live. To ensure that it is not possible to get a shock, safety precautions must be taken. If the voltage on the equipment you are working on exceeds 50 V a.c. or 120 V d.c., special test probes must be used (Figure 8.15). All test probes should meet the requirements of the Health and Safety Executive Guidance Note GS38.

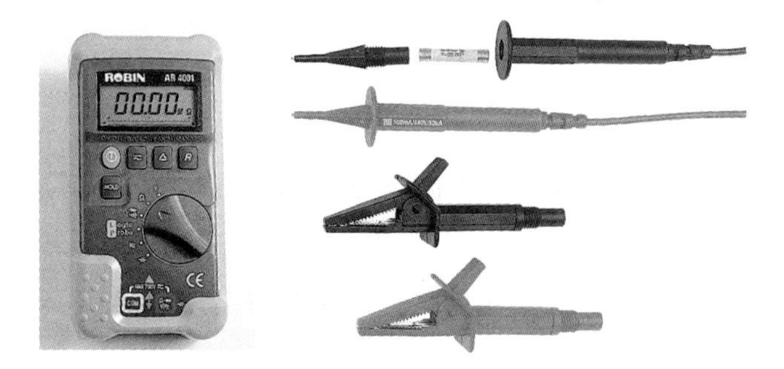


*Figure 8.15* *Fused test prods*

Voltage can be measured across a component or between any two connections in a circuit (Figure 8.16). As with all testing it is important to have some idea what the reading should be. If this is not possible the meter should be switched to a high range and then brought down to a suitable range when you can see which range is appropriate.

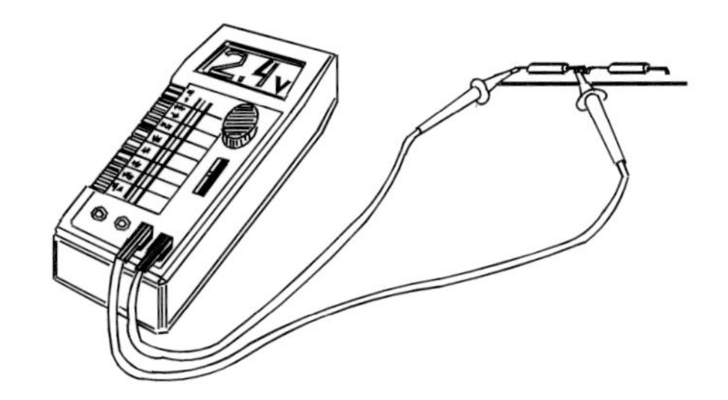

*Figure 8.16* *Measuring voltage drop*

### *Remember*

**The instrument used for measuring voltage may be the same one as used for resistance but switched to the voltage range.**

### *Try this*

## Measuring voltage

Write the reading of the meters in Figures 8.17 and 8.18 in the spaces provided. The range switch setting shows the maximum that can be indicated, and the reading must be adapted accordingly.

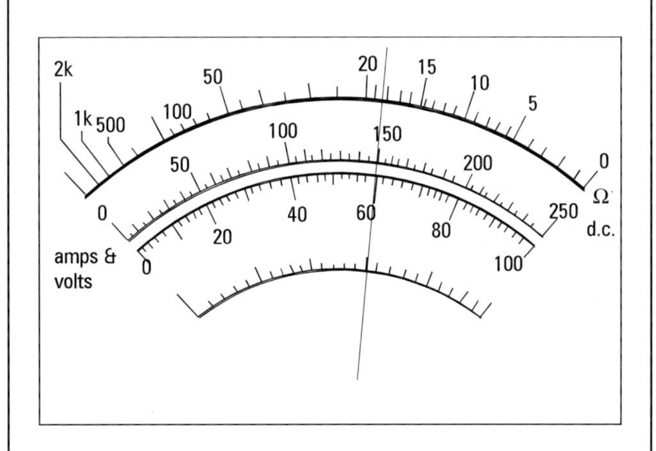


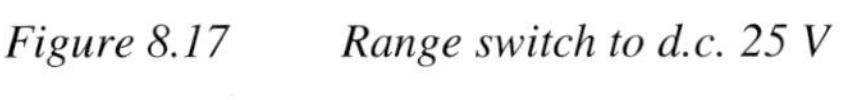

*Figure 8.17* *Range switch to d.c. 25 V*

Reading: ________

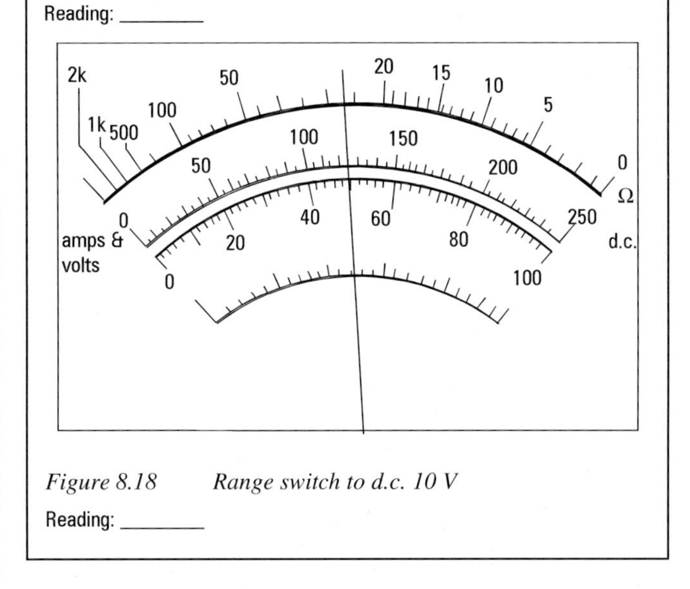


*Figure 8.18* *Range switch to d.c. 10 V*

Reading: ________

## Measuring current

When an instrument is connected to measure current (Figure 8.19) it should be in series with the load, so that the full current flows through the meter. It is important that the instrument is suitable for the current that is being measured.

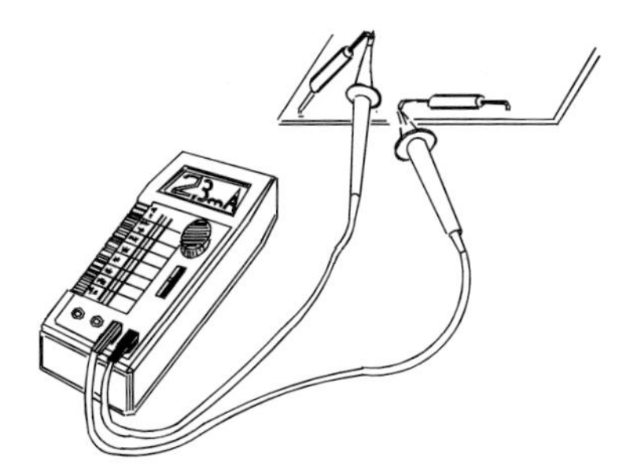

*Figure 8.19* *Measuring current*

The supply should be switched off when the meter is connected and disconnected. Often a circuit has to be broken to connect the meter in series with the load. The circuit should be reconnected after the readings have been completed.

In Figure 8.19 the two resistors normally connected in series have been separated and the ammeter connected between them.

**Always connect an ammeter into the circuit when the supply is switched off.**

*Try this*

## Measuring current

Write the reading of the meters in Figures 8.20 and 8.21 in the spaces provided. The range switch setting shows the maximum that can be indicated and the reading must be adapted accordingly

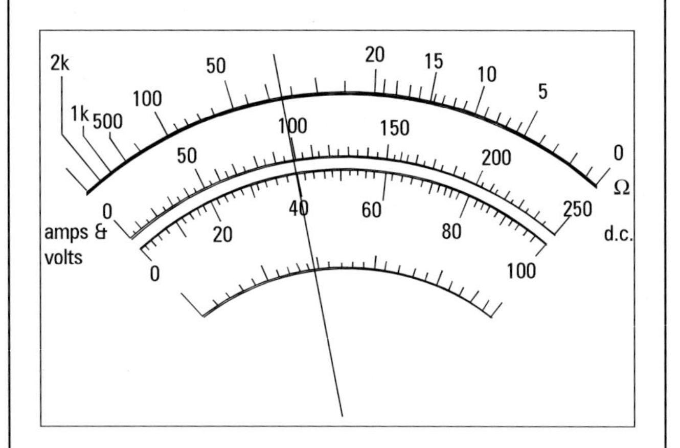


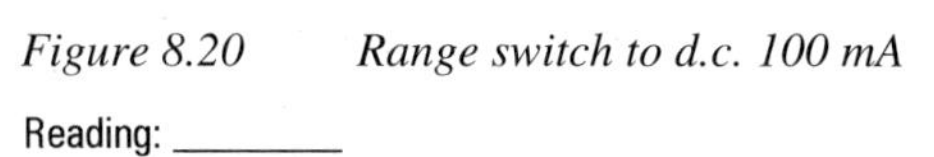

*Figure 8.20* *Range switch to d.c. 100 mA*

Reading: ________

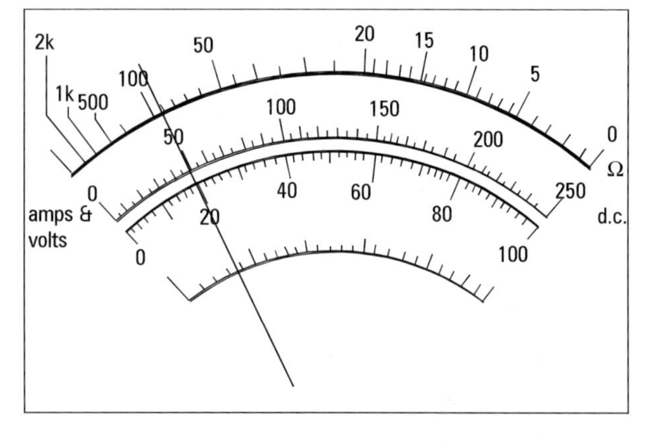


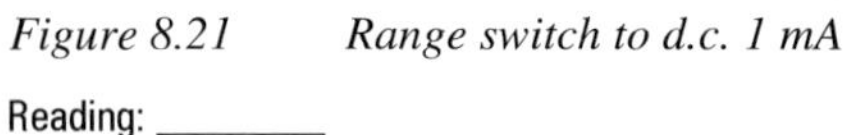

*Figure 8.21* *Range switch to d.c. 1 mA*

Reading: ________

# Wave form

## R.M.S. and peak value

Capacitors always have a voltage rating, and if that is exceeded, either in use or in testing, the dielectric may be damaged. The voltage rating refers to the maximum or peak voltage (amplitude) that may be applied. With an a.c. supply this is greater than the voltage usually referred to.

An a.c. mains supply of 230 V is not the maximum voltage that is applied. It is a value of voltage that will give an equivalent amount of power to a 230 V d.c. supply.

The maximum or peak voltage is about 325.22 V (Figure 8.22) and this is the minimum value which capacitors must be rated at.

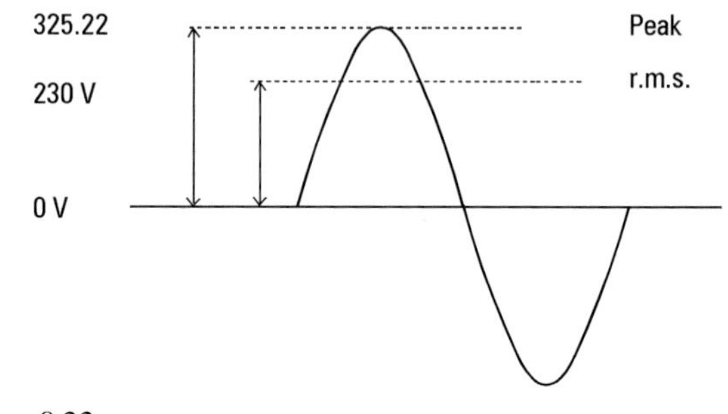


*Figure 8.22*

The r.m.s. (root mean square) value, as the working voltage is known as, is 0.707 of the peak value (Figure 8.23). Or in other words the peak value is 1.414 times the r.m.s. value.

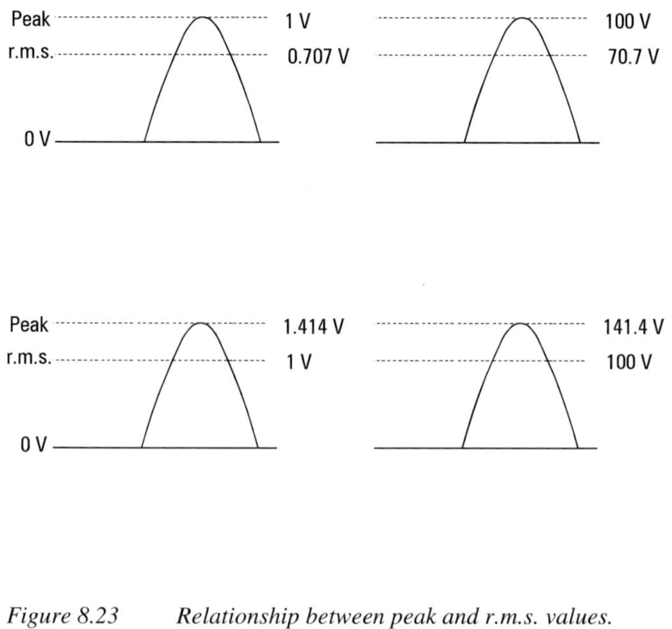


*Figure 8.23* *Relationship between peak and r.m.s. values. These are shown over half a cycle for clarity.*

## Average values

Over one complete cycle of 360° the average value is zero. This is calculated by taking the area of the positive half-cycle and taking away the area of the negative half. As these are the same but in opposite directions (Figure 8.24) they cancel each other out.

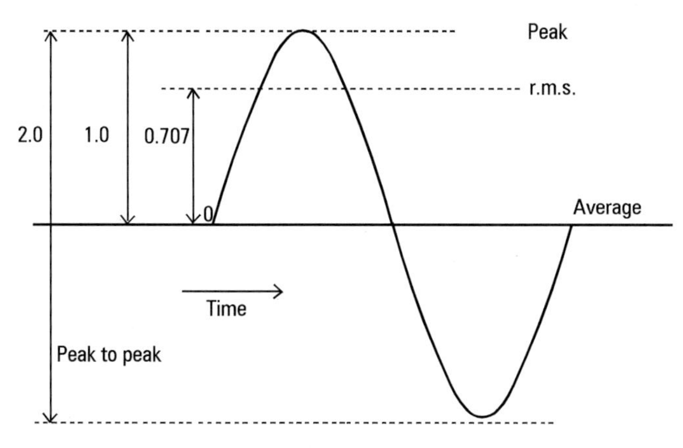


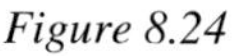
*Figure 8.24*

It is occasionally necessary to know the average value of one half-cycle. As the shape is not symmetrical the average is not 0.5 of the peak – it is in fact 0.637 (Figure 8.25).

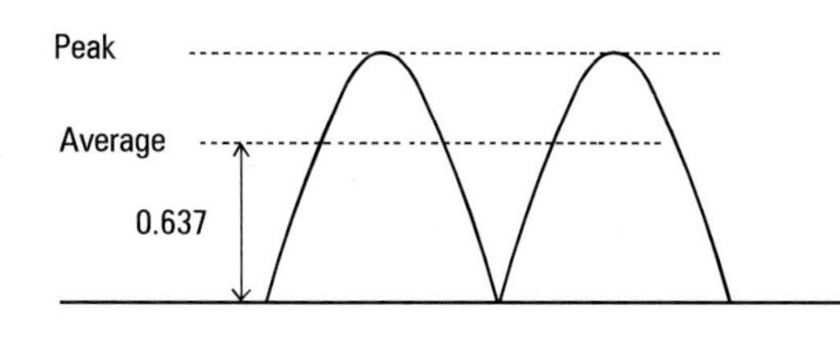


*Figure 8.25* *Average over half a cycle*

## Square waves

The supplies to some electronic components are not a.c. sine waves but square waveforms (Figure 8.26).

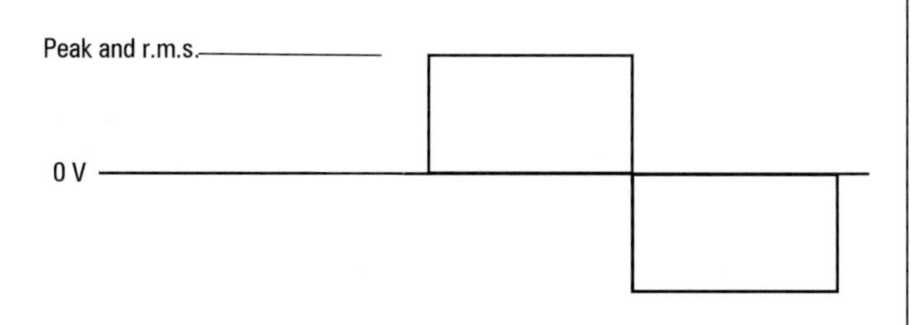


*Figure 8.26* *Square wave form*

When a square waveform is used the peak and working voltages (r.m.s.) are the same.

## Pulse waveform

A pulse waveform (Figure 8.27) is a sequence of on/off periods.

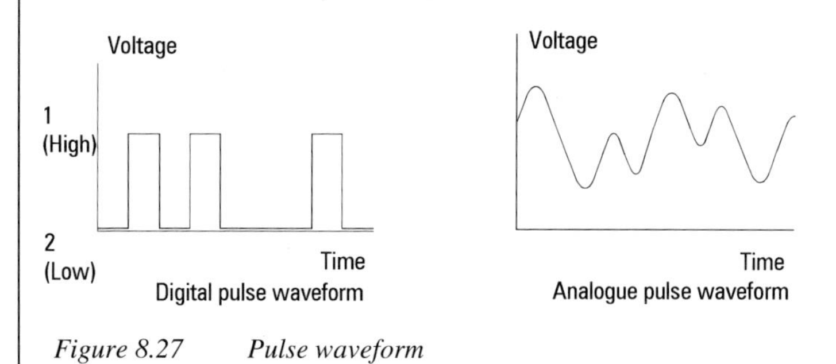


*Figure 8.27* *Pulse waveform*

## Frequency

Each cycle is completed in 360° (Figure 8.28).

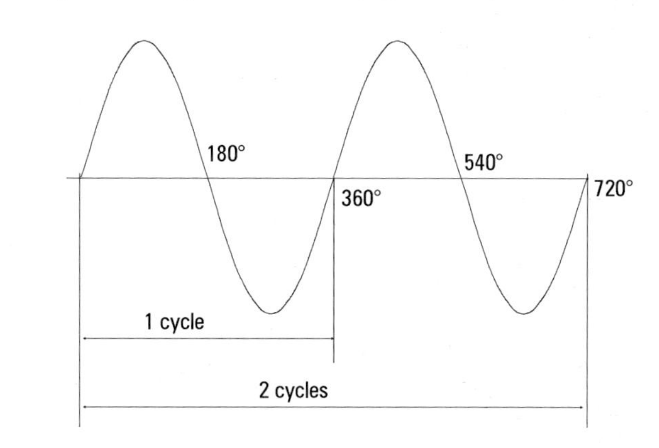


*Figure 8.28*

Each complete cycle has one positive and one negative half-cycle.

In some circuits it is important to plot more than one frequency on the same axis (Figure 8.29). In these cases it may be necessary to work out the frequency of each waveform and the phase relationship of one to the other.

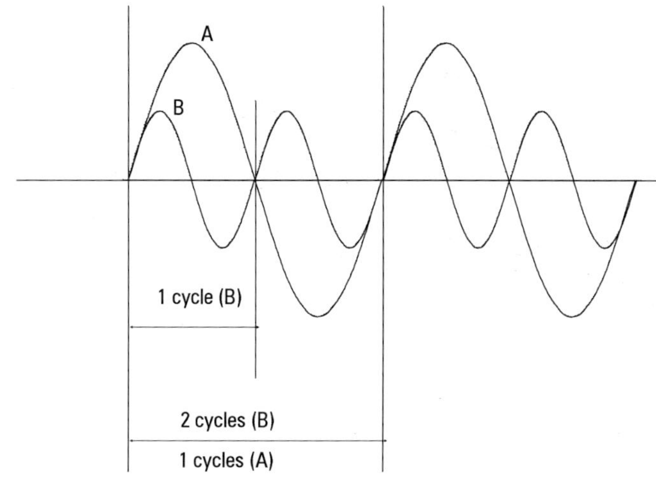

*Figure 8.29*

The frequency of waveform A is half that of B.

The time (symbol $T$) taken to complete one cycle is called the "period". The number of cycles completed in a time of one second is called the frequency (symbol $f$).

The relationship between these is shown by

$$f = \frac{1}{T}$$

or

$$T = \frac{1}{f}$$

## The cathode ray oscilloscope (CRO)

Waveform measurements would be carried out on an oscilloscope (Figure 8.30).

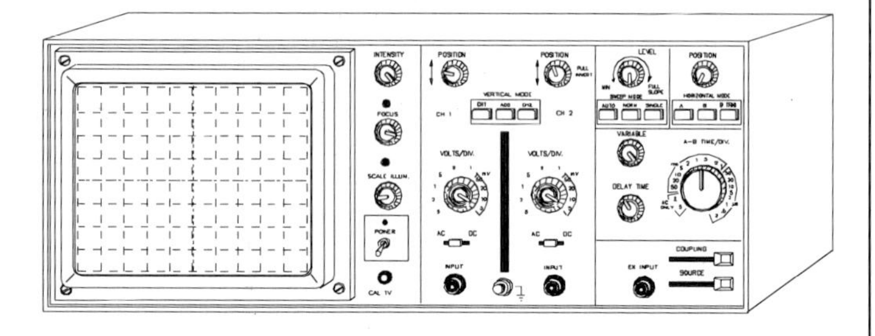

*Figure 8.30* *The cathode ray oscilloscope (CRO)*

The business end of the cathode ray oscilloscope (CRO) is the cathode ray tube. This performs the same function as it does in a TV set, in as much as it displays a picture. The screen is divided up into one centimetre squares and this pattern is known as the graticule. The graticule is centred on the intersection of the two axes.

The X-axis is horizontal and is normally used as a time scale. The time scale is variable and the calibration can be selected using the time base selector switch, which has a range of settings from seconds per centimetre, through milliseconds per centimetre, to microseconds per centimetre.

The trace on the CRO screen is actually the path taken by a spot of light on the screen. With no signal voltage connected and the time base calibration on its lowest setting you will see the spot coming on to the screen on the left and disappearing out at the right-hand edge, only to appear again on the left. As you speed up the time scale the dots move faster until they merge to form a continuous line.

The Y-axis is vertical and this also has a calibrated setting in volts per centimetre on the Y, or amplitude, control. The sensitivity can be adjusted right down to a few microvolts per centimetre.

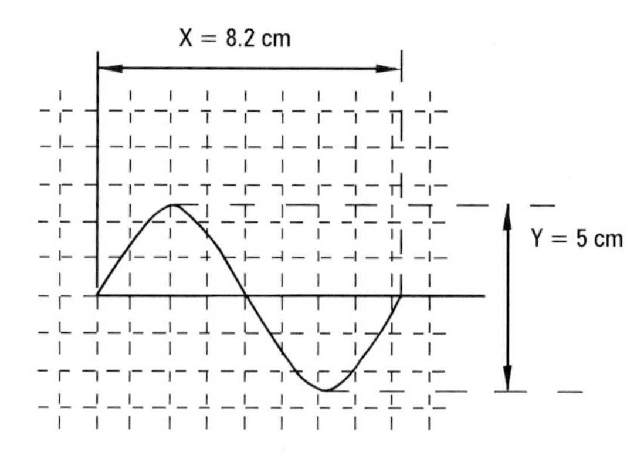

*Figure 8.31*

The trace on the screen in Figure 8.31 shows a sine wave with the graticule superimposed on it. From this, and the settings indicated on the controls, we will try to find out all we can about the sine wave displayed. From the trace we see that the length of the sine wave is 8.2 cm and the amplitude or peak-to-peak height is 5 cm.

## Time base

The setting on the X control is 2 milliseconds per centimetre. Therefore the time taken to complete one complete cycle is 8.2 cm at 2 ms per centimetre

$\therefore$ the period of the wave is $0.002 \times 8.2 = 0.0164$ seconds

The frequency is the reciprocal of the period

$$f = \frac{1}{0.0164}$$

$$f = 60.97 \text{ Hz}$$

**Note**

This type of measurement does not warrant such accuracy so the result would probably be taken as 61 Hz.

## Amplitude

The overall peak-to-peak voltage (Figure 8.32) is 5 cm at 1 volt/cm = 5 V.

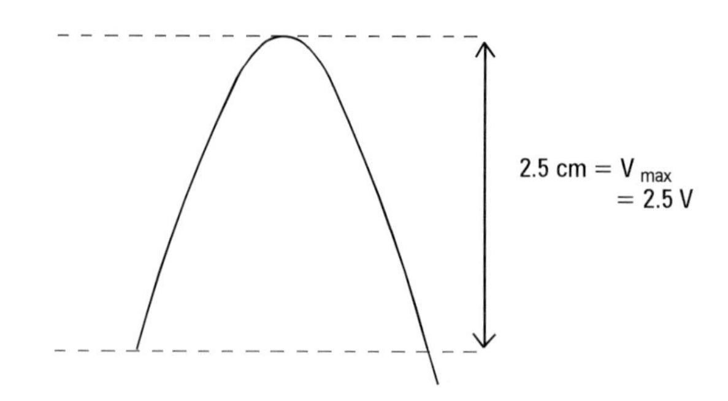

*Figure 8.32*

$\therefore$ the peak value on a half-cycle is 2.5 V.

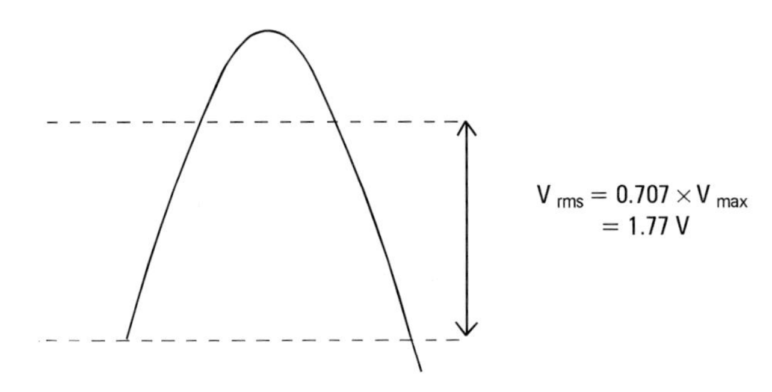

*Figure 8.33*

The r.m.s. will be 0.707 of the peak value, i.e. $V$ (r.m.s.) = 1.77 V (Figure 8.33).

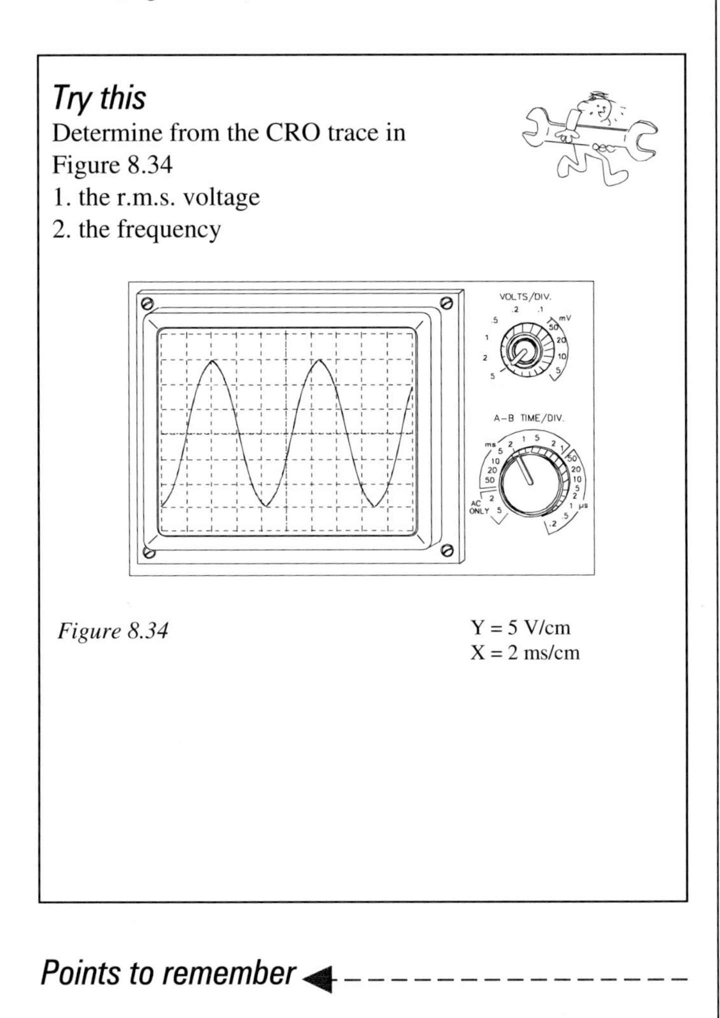

## *Try this*

Determine from the CRO trace in Figure 8.34

1. the r.m.s. voltage
2. the frequency



*Figure 8.34*

Y = 5 V/cm
X = 2 ms/cm

## *Points to remember*

When testing diodes it is important to check on the polarity of the ohmmeter. Why?

When connecting a measuring instrument into a circuit to measure current it must always be connected in _________ with the load. When the meter is connected or disconnected the supply should be switched off.

What is the relationship between r.m.s. voltage and peak voltage?

## *Self-assessment multi-choice questions*

**Circle the correct answers in the grid below.**

1. (i) Resistance measurements are always taken with the supply to the circuits disconnected.
   (ii) The ohmmeter is powered from its own internal supply.
   (a) only statement (i) is correct
   (b) only statement (ii) is correct
   (c) both statements are correct
   (d) neither statement is correct
2. Voltage measurements are taken
   (a) across components
   (b) in series with components
   (c) with the circuit supply disconnected
   (d) using the internal meter supply
3. Current measurements are taken
   (a) across components
   (b) in series with components
   (c) with the circuit supply disconnected
   (d) using the internal meter supply
4. The r.m.s. voltage is found to be 200 V. This means that the peak voltage must be approximately
   (a) 200 V
   (b) 282 V
   (c) 127 V
   (d) 141 V
5. The waveform in Figure 8.35 has a grid over it in which each vertical section is equal to 25 V. The peak-to-peak voltage for the waveform is

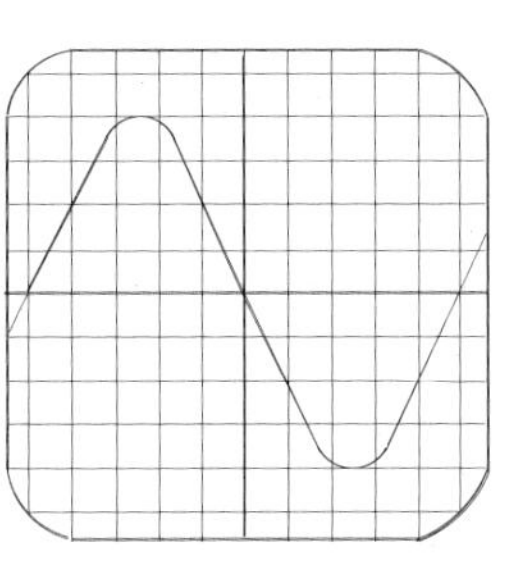

*Figure 8.35*

   (a) 100 V
   (b) 200 V
   (c) 125 V
   (d) 250 V

## *Answer grid*

| | | | | |
|---|---|---|---|---|
| 1 | a | b | c | d |
| 2 | a | b | c | d |
| 3 | a | b | c | d |
| 4 | a | b | c | d |
| 5 | a | b | c | d |

# End test

**Circle the correct answers in the grid at the end of the multi-choice questions.**

1. What is the length of the perimeter of a circle that has a radius of 3 cm?
   (a) 9.43 cm
   (b) 18.86 cm
   (c) 28.29 cm
   (d) 1.05 cm
2. Volume is measured in the unit
   (a) metre per second
   (b) ohm metre
   (c) square metre
   (d) cubic metre
3. The output power of a motor is 9 kW. If the percentage efficiency of the motor is 75%, what was the input power?
   (a) 6.75 kW
   (b) 10.25 kW
   (c) 12 kW
   (d) 15 kW
4. A conductor has a diameter of 4 mm and is 12 m long. A similar conductor is 3 m long. What would the diameter have to be if the resistance stays the same?
   (a) 1 mm
   (b) 2 mm
   (c) 3 mm
   (d) 4 mm
5. The quantity symbol for the resistivity of a material is
   (a) $\rho$
   (b) $\pi$
   (c) $R$
   (d) $\alpha$
6. How many coulombs of electricity are used when 13 amperes flow for 1 hour?
   (a) 276.92 coulombs
   (b) 780 coulombs
   (c) 7800 coulombs
   (d) 46 800 coulombs
7. The current taken by a 3 kW heater supplied with 240 V is
   (a) 0.08 A
   (b) 12.5 A
   (c) 80 A
   (d) 720 A
8. When measuring capacitance microfarads are the quantity usually used. What value does the micro ($\mu$) have?
   (a) $10^6$
   (b) $10^3$
   (c) $10^{-3}$
   (d) $10^{-6}$
9. A transformer has 150 turns on the input side and 600 turns on the output side. If the output voltage is 200 V what is the input voltage?

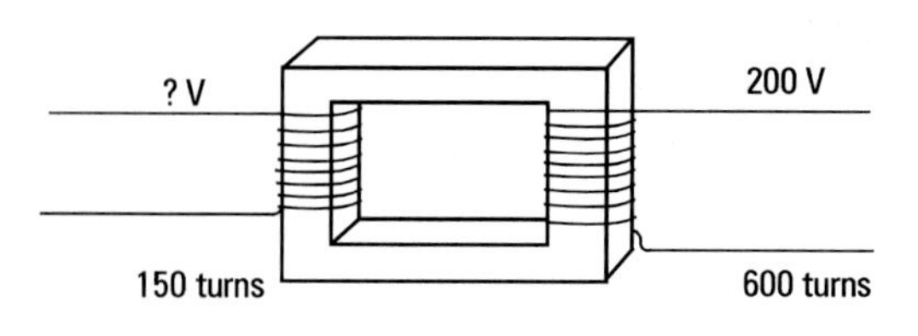


   (a) 800 V
   (b) 400 V
   (c) 100 V
   (d) 50 V
10. The output of a star connected distribution transformer winding has 415 V between phases. Phase to neutral voltage is
   (a) 115 V
   (b) 207.5 V
   (c) 240 V
   (d) 718 V
11. The capacitor in a fluorescent fitting is to
   (a) create a high inductance
   (b) create a higher current in the circuit conductors
   (c) step down the voltage
   (d) improve the power factor
12. The electrolyte in a lead acid cell is a mixture of
   (a) water and lead
   (b) mercuric acid and water
   (c) lead and acid
   (d) sulphuric acid and water
13. If the electrolyte comes into contact with the skin you must
   (a) leave it to dry
   (b) wash with plenty of clean water
   (c) soak in oil
   (d) wrap in lint
14. The most appropriate voltage to recharge a 12 V car battery would be
   (a) 15 volts a.c.
   (b) 15 volts d.c.
   (c) 110 volts a.c.
   (d) 230 volts a.c.
15. The type of battery used in modern motor cars gives a rated voltage of
   (a) 12 V
   (b) 240 V
   (c) 6 V
   (d) 90 V

16. If a lead acid battery is discharged the electrolyte will be
   (a) water with weak acid
   (b) high in acid content
   (c) high in zinc content
   (d) lead sulphate
17. Two resistors each of 50 Ω are connected in parallel to a 12 V d.c. supply. If they were to be replaced with a single resistor of an equivalent value this would have to be
   (a) 10 Ω
   (b) 50 Ω
   (c) 25 Ω
   (d) 4K166
18. The value of the resistor shown would be
   (a) 10 kΩ
   (b) 1 kΩ

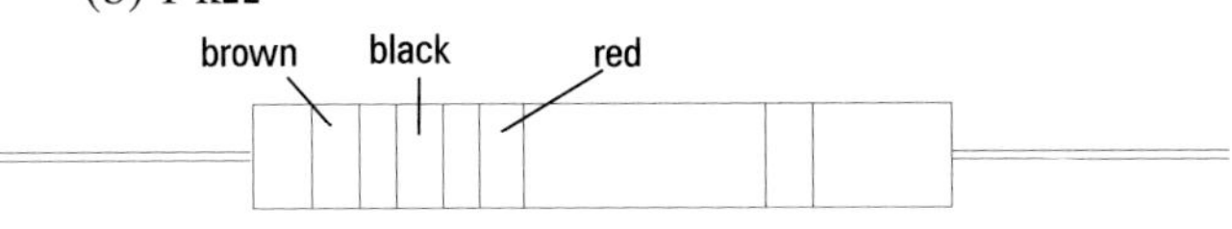


   (c) 102Ω
   (d) 100Ω
19. The voltage drop across a 150 Ω resistor supplying a current of 1 mA is
   (a) 0.15 V
   (b) 1.5 V
   (c) 15 V
   (d) 150 V
20. The resistor shown could have a resistance between

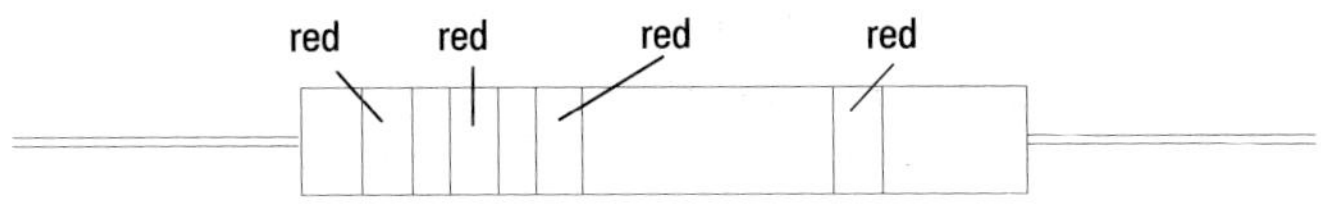


   (a) 2156 and 2244 Ω
   (b) 2090 and 2130 Ω
   (c) 1980 and 2420 Ω
   (d) 1760 and 2640 Ω
21. Two factors that can decrease the capacitance of a capacitor are
   (a) smaller plates, thicker dielectric
   (b) smaller plates, thinner dielectric
   (c) larger plates, thicker dielectric
   (d) larger plates, thinner dielectric
22. (i) The charge on a capacitor can be calculated from $Q = VC$.
   (ii) When two capacitors of equal value are connected in series their combined value is half that of each individual capacitor.
   (a) only statement (i) is correct
   (b) only statement (ii) is correct
   (c) both statements are correct
   (d) neither statement is correct
23. 1μF is equal to
   (a) 10 nF
   (b) 100 pF
   (c) 1000 pF
   (d) 1000 nF
24. The symbol shown could be used to represent
   (a) a polarised capacitor
   (b) a plastic film capacitor
   (c) a polystyrene capacitor
   (d) a silver mica capacitor
25. A low frequency transformer has a turns ratio of 5:1. If the output voltage is 48 V the input voltage will be
   (a) 240 V
   (b) 53 V
   (c) 43 V
   (d) 9.6 V
26. The symbol shown represents a
   (a) variable resistor
   (b) preset inductor
   (c) double wound transformer
   (d) variable inductor
27. Two factors that can increase the inductance of a coil are
   (a) more turns on the coil, reduced iron core in the coil
   (b) less turns on the coil, reduced iron core in the coil
   (c) more turns on the coil, increased iron core in the coil
   (d) less turns on the coil, increased iron core in the coil
28. (i) An inductor works on the principle of self inductance
   (ii) A transformer works on the principle of mutual inductance
   (a) only statement (i) is correct
   (b) only statement (ii) is correct
   (c) both statements are correct
   (d) neither statement is correct
29. The electronic component shown is
   (a) a light emitting diode
   (b) a silicon controlled rectifier
   (c) an integrated circuit
   (d) a photodiode
30. The output (V) of the circuit shown will be

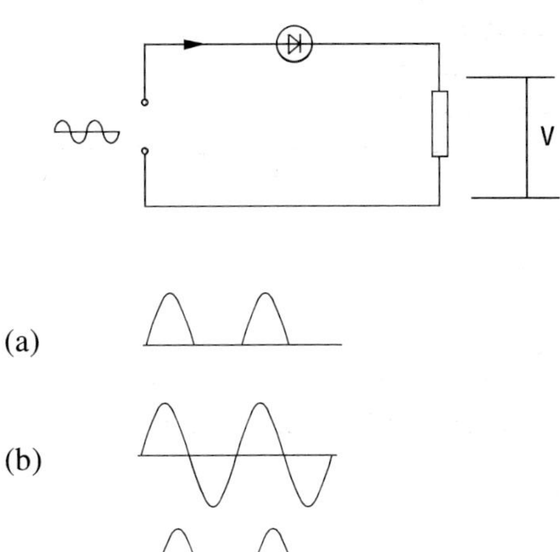


   (a)
   (b)
   (c)
   (d)

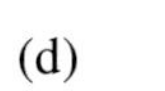

31. The graphical symbol shown represents a
 (a) thyristor
 (b) triac
 (c) transistor
 (d) light-emitting diode

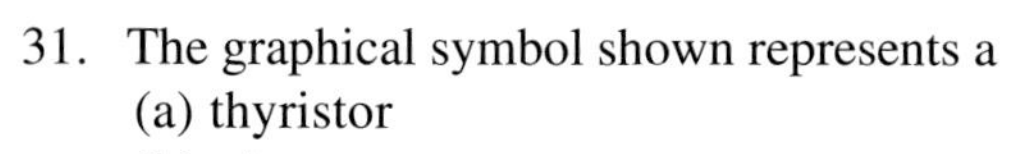
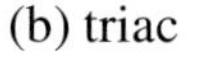

32. An example of a semiconductor material is
 (a) copper
 (b) aluminium
 (c) germanium
 (d) carbon

33. The block diagram symbol shown represents

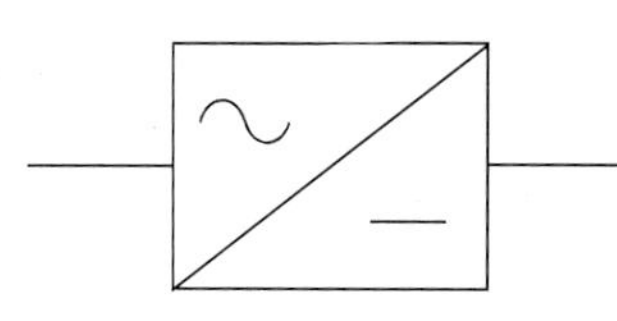

 (a) an amplifier
 (b) a transformer
 (c) a rectifier
 (d) a power supply

34. The graphical symbol shown represents a
 (a) variable inductor
 (b) preset capacitor
 (c) polarised capacitor
 (d) variable capacitor

35. The device which is effectively two thyristors back to back is
 (a) a triac
 (b) an LED
 (c) a thermistor
 (d) a diode

36. An AND gate for inputs A and B would give an output when
 (a) no signal present at A and B
 (b) signal present at A and B
 (c) signal present at A, not at B
 (d) signal present at B, not at A

37. The type of plug shown is called a
 (a) coaxial
 (b) jack
 (c) module
 (d) DIN

38. The solder used for connecting electronic components is an alloy generally made up of
 (a) 60% tin, 40% lead
 (b) 40% tin, 60% lead
 (c) 40% tin, 60% copper
 (d) 60% copper, 40% tin

39. The r.m.s. voltage is found to be 210 V. This means that the peak voltage must be approximately
 (a) 210 V
 (b) 297 V
 (c) 105 V
 (d) 330 V

40 The waveform below has a grid over it in which each vertical section is equal to 20 V. The peak-to-peak voltage for the waveform is

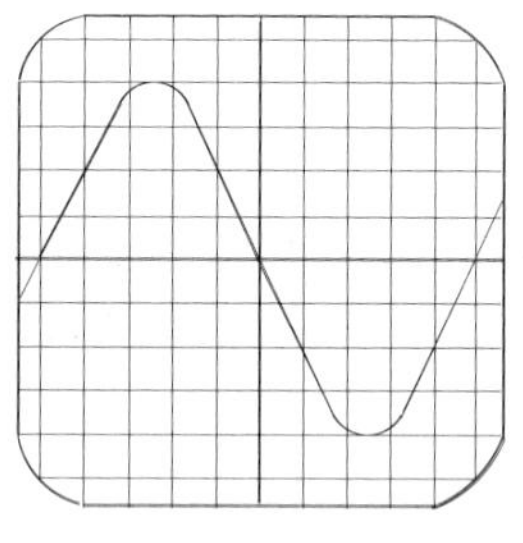

 (a) 80 V
 (b) 160 V
 (c) 100 V
 (d) 200 V

## *Answer grid*

| | | | | | | | | | |
|---|---|---|---|---|---|---|---|---|---|
| 1 | a | b | c | d | 21 | a | b | c | d |
| 2 | a | b | c | d | 22 | a | b | c | d |
| 3 | a | b | c | d | 23 | a | b | c | d |
| 4 | a | b | c | d | 24 | a | b | c | d |
| 5 | a | b | c | d | 25 | a | b | c | d |
| 6 | a | b | c | d | 26 | a | b | c | d |
| 7 | a | b | c | d | 27 | a | b | c | d |
| 8 | a | b | c | d | 28 | a | b | c | d |
| 9 | a | b | c | d | 29 | a | b | c | d |
| 10 | a | b | c | d | 30 | a | b | c | d |
| 11 | a | b | c | d | 31 | a | b | c | d |
| 12 | a | b | c | d | 32 | a | b | c | d |
| 13 | a | b | c | d | 33 | a | b | c | d |
| 14 | a | b | c | d | 34 | a | b | c | d |
| 15 | a | b | c | d | 35 | a | b | c | d |
| 16 | a | b | c | d | 36 | a | b | c | d |
| 17 | a | b | c | d | 37 | a | b | c | d |
| 18 | a | b | c | d | 38 | a | b | c | d |
| 19 | a | b | c | d | 39 | a | b | c | d |
| 20 | a | b | c | d | 40 | a | b | c | d |

# Answers

**These answers are given for guidance and are not necessarily the only possible solutions.**

## Chapter 1

p.1 Ampère, Newton, Joule, Watt, Pascal, Volt, Ohm, Siemens

p.3 Try this: mm = millimetre, mA = milliampere, GW = gigawatt, kV = kilovolt, mΩ = milliohm, pF = picofarad, μC = microcoulombs

p.4 Try this: (1) 1875 $mm^2$; (2) 9.62 $mm^2$
Try this: (1) 24 cm; (2) 22 cm; (3) 44 cm

p.5 Try this: 15 cm

p.6 Try this: (1) 9 $m^3$; (2) 0.29 $m^3$

p.7 Try this: 35.71 $m^3$
(cube 64 $m^3$ minus the cylinder 28.29 $m^3$)

p.8 Try this: (1) 6, 6.5; (2) 400 turns; (3) 76%; (4) 80%

p.9 (1) 3.775 Ω; (2) 2.372 Ω

p.11 (1) 20 V; (2) 3 A

p.12 Try this: $\rho = 17.8\ \mu\Omega = 17.8 \times 10^{-6} \times 10^{-3}\ \Omega$ m, $l = 100$ m, $A = 120\ \text{mm}^2 = 120 \times 10^{-6}\ \text{m}^2$

$$R = \frac{17.8 \times 10^{-6} \times 10^{-3} \times 100}{120 \times 10^{-6}}$$

= 0.015 Ω (or 15 mΩ)

p.13 Try this: (1) 85.06; (2) 0.027; (3) 0.36; (4) 0.027; (5) 17.9; (6) 104.47; (7) 893; (8) 12.38; (9) 0.0078; (10) 0.027; (11) 113.5; (12) 166; (13) 0.019; (14) 30.94; (15) 0.17

p.14 Try this: (1) 18.87; (2) 450; (3) 12.36 mm; (4) 79%; (5) 3.16 A
SAQ (1) c; (2) b; (3) c; (4) d; (5) a

## Chapter 2

p.15 metre: length; newton: force; farad: capacitance; weber: magnetic flux; tesla: magnetic flux density; ohm: resistance; $10^3$ is ten to the power of three or 1000; $10^1$ is ten to the power of 1 or 10, $10^{-9}$ is ten to the power of –9 or 0.000 000 001.

p.17 Try this: (1) 10 $m/s^2$; (2) 784.8 N; (3) 3433 N

p.18 Try this: (1) approx. 6 N; (2) approx. 6 N

p.20 Try this: 234 kg

p.22 Try this: 136.25 N

p.23 Try this: (1) 117.72 N; (2) 73.58 W; (3) work done = 441 450 N; power required = 11 036.25 W

p.25 Try this: $\alpha = \dfrac{\text{change of length}}{\text{original length} \times \text{temperature rise}}$

$$\text{temperature rise} = \frac{\text{change of length}}{\text{original length} \times \alpha}$$

$$\text{temperature rise} = \frac{0.02}{500 \times 11 \times 10^{-6}}$$

temperature rise = 36.36 °C
Add the temperature rise to the original temperature
36.36 + 10 = 46.36 °C

p.26 SAQ (1) c; (2) d; (3) d; (4) a; (5) d

## Chapter 3

p.27 (1) 1471.5 N; (2) 17.8 kg; (3) 48 N; (4) work done = 0.02 joules; power = 0.0044 W (4.44 mW)

p.32 Try this: 4 mm; 0.029 Ω (29 mΩ)

p.33 Try this: 88 Ω

p.34 Try this: (1) 0.071 Ω; (2) 67.2 Ω; (3) 15 000 C; (4) 15 A

p.38 Try this: (1) 9.2 A; (2) 5 A

p.39 Try this: (1) 24 Ω; (2) 3 Ω; (3) 88 V; (4) 1.44 A; (5) 48 Ω

p.40 Try this: $R_1 = 0.5\ \Omega$; $R_2 = 2\ \Omega$

p.42 Try this: (1) 2 A, 2 V, 6 V, 4 V; (2) 2.645 kW; (3) 0.65 A
SAQ (1) b; (2) c; (3) d; (4) b; (5) a

## Chapter 4

p.43 Good conductors: copper, aluminium, silver; good insulators: air, glass, ceramics
The length of the conductor: the longer the conductor the greater the resistance to current flow
The temperature of the conductor: the higher the temperature the greater the resistance to current flow
The cross-sectional area of the conductor: the thinner the conductor the greater the resistance to current flow
The three main effects of current flow are the production of heat, the chemical effect and the production of magnetism
The relationship of voltage, current and resistance is given in the formula called Ohm's Law: $V = I R$

p.45 Try this: (1) alternating current; (2) to connect the coil to the meter; (3) 360°

p.47 Try this

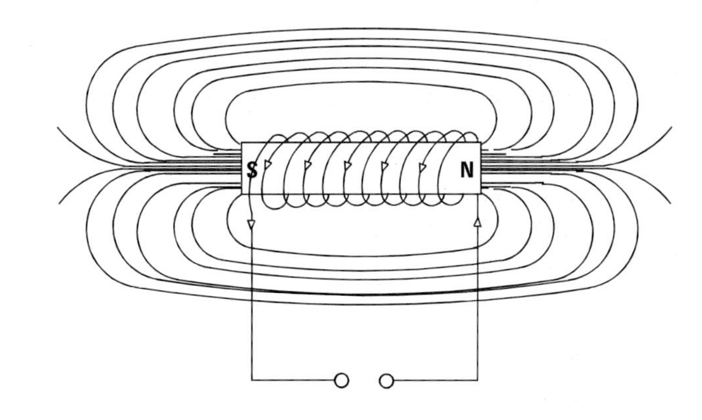


p.48 Try this: 30 V

p.49 Try this: 15 A

p.55 Try this: (1) a. 3; b. 12

p.58 SAQ (1) d; (2) b; (3) c; (4) a; (5) c

pp. 59 and 60: Progress check

(1) a; (2) b; (3) d; (4) a; (5) c; (6) c; (7) b; (8) b; (9) c; (10) a; (11) c; (12) d; (13) a; (14) d; (15) d; (16) b; (17) a; (18) c; (19) b; (20) a; (21) d; (22) c; (23) a; (24) d; (25) d

## Chapter 5

p.61 Mechanical; conductor and magnetic field; chemical
Primary cells produce electricity from a chemical reaction in the cell when connected to a circuit, and secondary cells use a chemical action but in effect store electricity in a chemical form. In a primary cell, when the chemical action stops the battery is exhausted, but the secondary cell can be recharged many times over.

p.63 Try this: 100 Ω

p.68 SAQ (1) d; (2) a; (3) a; (4) c; (5) c

p.71 Try this: (1) 51 000 Ω; (2) 2400 Ω; (3) 27 Ω; (4) 390 Ω; (5) 680 000 Ω

p.72 Try this: (1) a. 44650 Ω; b. 49350 Ω; (2) a. 24.3 Ω; b. 29.7 Ω; (3) a. 1710 Ω; b. 1890 Ω; (4) a. 323.4 Ω; b. 336.6 Ω; (5) a. 17600 Ω; b. 26400 Ω

p.73 (1) 10 Ω ± 5%; (2) 15 kΩ ± 10%; (3) 1 MΩ ± 20%; (4) 1222 kΩ ± 10%; (5) 680 Ω ± 20%; (6) 1M5J; (7) R47G; (8) 10MF; (9) 330K; (10) 1800KG
SAQ (1) d (2) a (3) d (4) c (5) b

p.81 SAQ (1) d (2) c (3) b (4) c (5) a

p.85 SAQ (1) b (2) b (3) c (4) c (5) a

p.89 Try this: (1) bridge rectifier; (2) LED

p.92 SAQ (1) c; (2) a; (3) d; (4) d; (5) b

## Chapter 6

p.93 potentiometer; temperature measurement and control etc.; the presence of a colour band or number code; microfarads (μF); aluminium can; indentation or coloured band; three terminals – emitter, base and collector; two main types of transistor are pnp and npn.

p.101

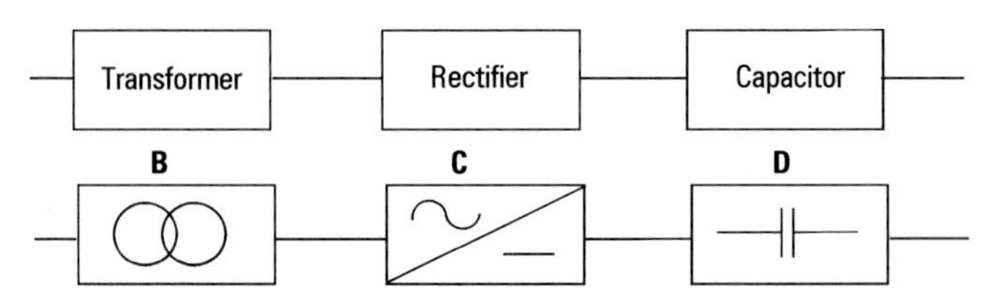


*Block diagram*

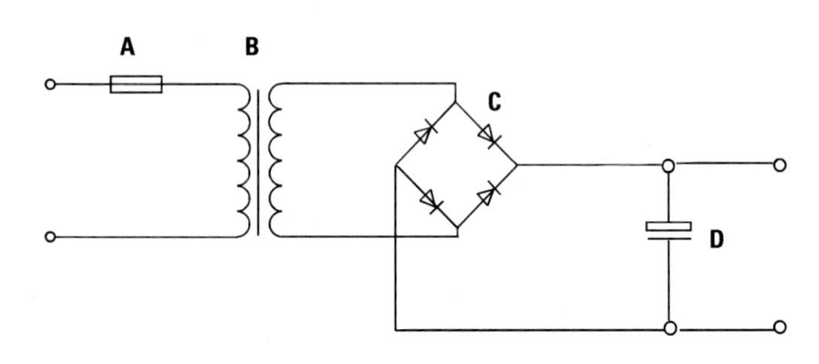


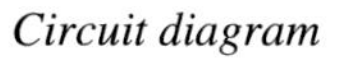

*Circuit diagram*

p.102

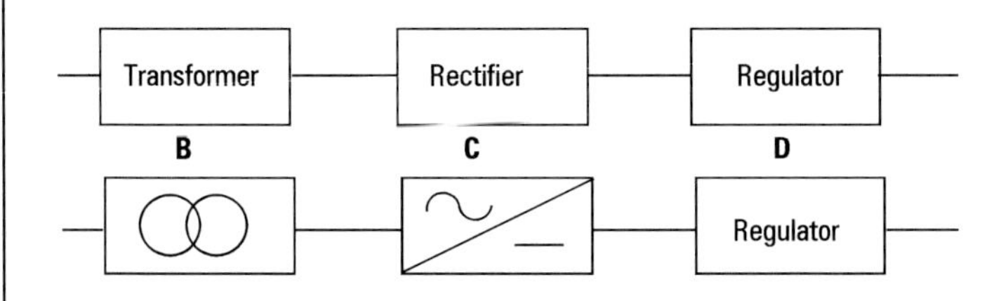


*Block diagram*

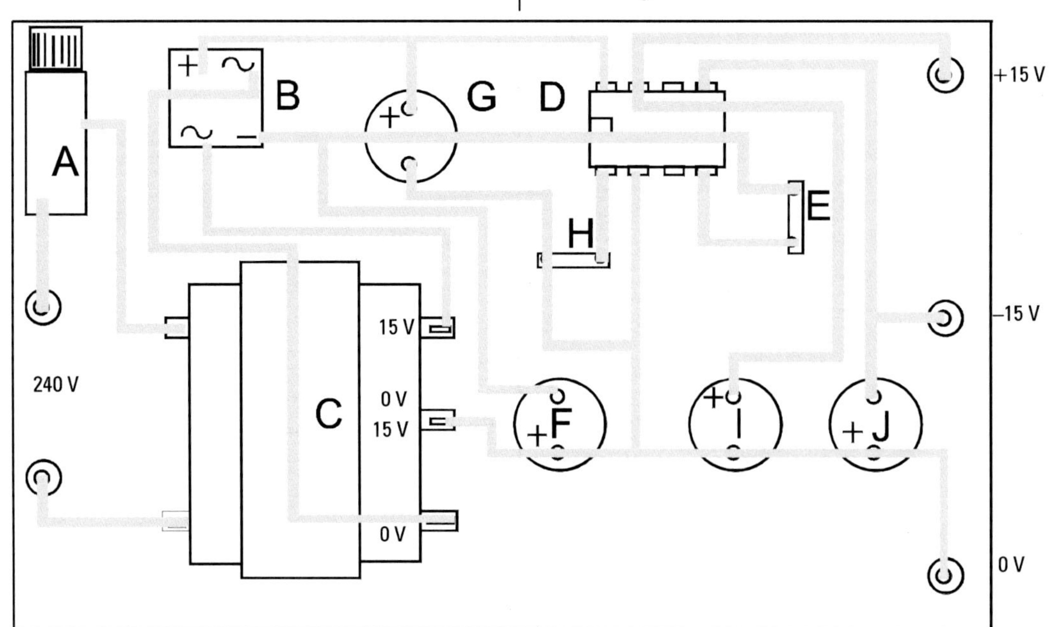


*Layout diagram*

p.104 SAQ (1) b; (2) d; (3) b; (4) a; (5) d

## Chapter 7

p.105 Block; circuit; layout; logic; component positional reference; scaled representation of the actual objects

p.112 SAQ (1) c; (2) a; (3) c; (4) b; (5) c

## Chapter 8

p.113 Plugs and sockets; connector blocks; insulation displacement connections or crimped connections; gold; tools required: soldering iron, round nosed pliers, lead forming tool; 60% tin, 40% lead solder; dry joints; high-resistance joints; excess solder and too much heat

p.114 Try this: 35 Ω, 10 kΩ

p.115 Try this: 6 Ω; > 2 kΩ; the diode is working correctly

p.116 Try this: 100 Ω; 2 Ω; the transistor is not working correctly

p.117 Try this: 15 V; 5 V

p.118 Try this: 40 mA; 0.2 mA

p.121 Try this: (1) 10.6 V; (2) 119 Hz
SAQ (1) c; (2) a; (3) b; (4) b; (5) b

pp.122–124 End test
(1) a; (2) d; (3) c; (4) b; (5) a; (6) b; (7) b; (8) d; (9) a; (10) c; (11) d; (12) d; (13) b; (14) b; (15) a; (16) a; (17) c; (18) b; (19) a; (20) a; (21) a; (22) b; (23) d; (24) a; (25) a; (26) d; (27) c; (28) c; (29) b; (30) a; (31) d; (32) c; (33) c; (34) d; (35) a; (36) b; (37) a; (38) a; (39) b; (40) b

# Appendix

## Terms used in this book

### Associated science

***acceleration***
an increase in velocity (p. 2)

***alternator***
a machine that converts mechanical energy into electrical energy in the form of alternating current (p. 43)

***area***
a length multiplied by a length measured in square metres, $m^2$ (p. 2)

***coefficient of linear expansion***
the expansion per unit length of material per unit temperature rise (p. 24)

***conductor***
a material that transmits electricity (p. 28)

***commutator***
a rotating switch for reversing an electric current (p. 52)

***current***
the movement of electrically charged particles (p. 28)

***density***
how compact a substance is (p. 20)

***distance***
the length of space between two points (Ch. 1)

***efficiency***
the ratio of useful energy output and the energy input (p. 8)

***e.m.f.***
the force that moves electrons (p. 33)

***energy***
the capacity to do work (p. 22)

***equilibrium***
when forces acting on a body are equal and opposite (p. 18)

***force***
an influence tending to cause the motion of a body (p. 15)

***frequency***
the number of cycles per second (p. 45)

***gravity***
the force exerted on a body by the earth (p. 16)

***lever***
a bar pivoted in order to rotate about a point (p. 20)

***magnetic flux***
the lines of force surrounding magnets and electric currents (p. 44)

***magnetic flux density***
the quantity of magnetic flux in a give area (p. 44)

***mass***
the quantity of matter that a body contains measured in kilograms, kg (p. 15)

***moment***
the turning effect of a force (p. 21)

***percentage***
the proportion per hundred (p. 8)

***potential***
the charge in an electrical field (p. 33)

***potential difference***
the difference in electrical potential between two points (p. 33)

***power***
the rate of working, electrical energy – found by the product of voltage and current (p. 23)

***primary winding***
the winding on a double-wound transformer connected to the supply (p. 47)

***resistance***
the opposition to a current in a circuit (p. 32)

***resistivity***
the resistivity of a material is the resistance of a sample of unit length and unit cross-sectional area (p. 32)

***secondary winding***
the winding on a double-wound transformer connected to the load (p. 47)

***shear force***
a sitiuation in which one part of a body is acted on by a force and forced sideways in relation to the other part of the body (p. 19)

***solenoid***
a coil of wire capable of carrying a current acting as a magnet (p. 46)

***speed***
the rate of motion (p. 2)

***stability***
a state of not being easily moved (p. 22)

***thermostat***
a device that automatically regulates temperature (p. 25)

***temperature***
the measure of heat in a body (p. 24)

***vector***
a quantity having both direction and magnitude (p. 17)

***volume***
the amount of space of a three-dimensional object or the space that a gas or liquid occupies (p. 6)

***weight***
the force experienced by a body due to the earth's gravitation (p. 16)

***work***
work is done whenever an object is moved by a force (p. 22)

## Electronics

***amplitude***
the maximum or peak value (p. 120)

***bipolar***
having two connections (p. 89)

***capacitance***
the ability to store an electrical charge (p. 74)

***capacitor***
a component capable of storing an electrical charge (p. 74)

***cathode ray oscilloscope***
a cathode ray oscilloscope displays waveforms on a screen (p. 120)

***circuit diagram***
a circuit diagram shows which components are connected together to show the operation of the circuit (p. 98)

***component positional reference system***
a circuit diagram on a grid so that accurate reference can be made to components and more detail given in a separate key (p. 98)

***crimp***
a compressed connection (p. 108)

***diac***
a diac is a two terminal connection which conducts in either direction (p. 90)

***DIL***
stands for "dual in-line" and refers to the connections of an integrated circuit (p. 67)

***diode***
a semiconductor device which allows the flow of current in one direction only (p. 87)

***edge connector***
a plug-in arrangement on the edge of a printed circuit board (p. 106)

***FET***
field-effect transistor (p. 90)

***flux***
a substance mixed with a metal to promote fusion when soldering (p. 109)

***inductor***
an inductor is a coil with a magnetic conductor through the centre (p. 84)

***insulation displacement connection***
a terminating technique which allows insulated conductors to be connected without first removing the insulation (p. 108)

***integrated circuit***
a complete electronic circuit formed on one piece of semiconductor material (p. 91)

***layout diagram***
a diagrammatic picture of how the finished circuit will appear using scaled shapes of the actual objects (p. 93)

***LED***
a light-emitting diode (p. 88)

***logic diagram***
shows how logic gates can be combined to provide a control function (pp. 93, 103)

***logic gate***
logic gates process digital input signals in order to give an output signal when certain input conditions are met (pp. 93, 103)

***plug and socket***
a form of connection used in electronics that is readily separable (p. 106)

***PN junction***
a kind of barrier between n-type and p-type material (p. 86)

***rectifier***
a device containing one or more diodes which is used to convert a.c. to d.c. (p. 90)

***relay***
a device which allows one circuit to operate another (p. 85)

***RF***
radio frequency (p. 84)

***resistor***
a device having resistance to an electric current (p. 62)

***single in-line***
stands for "single in-line", where all the connections of a circuit are in a single line (p. 67)

***sine wave***
a curve representing periodic oscillations (pp. 45, 118)

***solder***
a fusible alloy used to join metals or wires (p. 109)

***soldering iron***
a tool used to heat the metal or wire before soldering (p. 109)

***thermistor***
a temperature-sensitive resistor (p. 67)

***thyristor***
a type of silicon-controlled rectifier used in switching circuits (p. 90)

***transformer***
a basic transformer consists of two independent coils of wire connected only by a magnetic conductor and they rely on mutual inductance from one coil to another (p. 83)

***transistor***
a transistor is a three-terminal device capable of amplifying current (p. 89)

***triac***
a bi-directional switching device which can be used in a.c. circuits (p. 90)

***wire wrap***
a form of connection which is mechanically fixed (p. 107)

***zener***
a diode that provides a constant voltage (p. 88)

# Equations used in this book

For further information see the page indicated.

## Area

area of a rectangle (p. 3) = length × breadth

area of a circle (p. 3) $= \dfrac{\pi d^2}{4}$

area of a triangle (p. 3) $= \dfrac{1}{2}$ base × perpendicular height

## Perimeter

In a rectangle, triangle or other straight sided figure the perimeter is the addition of the lengths of the sides (p. 4). The perimeter of a circle is called the circumference (p. 4)

circumference $= \pi d$

## Pythagoras' theorem

(p. 5) $AC^2 = AB^2 + BC^2$

$\therefore \quad AC = \sqrt{AB^2 + BC^2}$

## Volume

The volume of a cuboid or rectangular prism (p. 6)

= length × breadth × depth

Volume of a cylinder (p. 6) $= \dfrac{\pi d^2 \times l}{4}$

Volume of a sphere (p. 7) $= \dfrac{4}{3}\pi r^3$

## Efficiency

Efficiency (p. 8) $= \dfrac{\text{output}}{\text{input}}$

Efficiency % (p. 8) $= \dfrac{\text{output}}{\text{input}} \times 100$

## Mechanical science

Force (p. 16) $= m \times a$

Density (p. 20) $= \dfrac{\text{mass}}{\text{volume}}$ or $= \dfrac{m}{V}$

Moment (p. 21) = force × distance or $= f_1 d_1$

Principle of moments (p. 21)

$f_1 d_1 = f_2 d_2$

Work done (joules) = force (newtons) × distance (metres)

or (p. 22) $W = Fd$

Power (p. 23) $= \dfrac{\text{work done}}{\text{time taken}}$

Coefficient of linear expansion, symbol $\alpha$ (p. 24)

$$\alpha = \frac{\text{change of length}}{\text{original length} \times \text{temperature change}}$$

## Electrical science

The resistance in a series circuit (pp. 9, 64)

$R_T = R_1 + R_2$

Ohm's Law (pp. 10, 38, 62, 63, 65)

$V = IR$

or $I = \dfrac{V}{R}$

or $R = \dfrac{V}{I}$

Resistivity, $\rho$ (pp. 10, 32)

$R = \dfrac{\rho l}{A}$

Power (pp. 13, 35, 40)

$P = I^2 R$

or $P = VI$

or $P = \dfrac{V^2}{R}$

Temperature coefficient of resistance, $\alpha$ (pp. 10, 32)

$$\frac{R_1}{R_2} = \frac{(1 + \alpha t_1)}{(1 + \alpha t_2)}$$

Quantity of electricity (p. 34)

amperes $= \dfrac{\text{quantity of electricity in coulombs}}{\text{time in seconds}}$

or $I = \dfrac{Q}{t}$ or $Q = It$

In a series circuit all of the currents $I$ will be the same value (p. 34) $I_T = I_A = I_B$

In a parallel circuit (p. 34)

$I_T = I_A + I_B$

Power factor (p. 51)

$P = VI \cos \phi$

## Electronics

The resistance in a parallel circuit (p. 65)

$$\frac{1}{R_T} = \frac{1}{R_1} + \frac{1}{R_2} + \frac{1}{R_3} + \frac{1}{R_4} \ldots \text{etc.}$$

Capacitance in a circuit

charge ($Q$) (p. 75) = voltage ($V$) × farads ($C$)

Energy stored in a capacitor ($W$) (p. 75) $= \dfrac{1}{2} C V^2$

For capacitors connected in parallel (p. 75)

$C_{\text{Total}} = C_1 + C_2 + C_3 + C_4 \ldots$ etc.

For capacitors connected in series (p. 76)

$$\frac{1}{C_{\text{Total}}} = \frac{1}{C_1} + \frac{1}{C_2} + \frac{1}{C_3} + \frac{1}{C_4} \ldots \text{etc.}$$

Inductance (p. 83), time constant (tau) $\tau = \dfrac{L}{R}$ seconds

where $L$ is the inductance in henrys and $R$ is the resistance in ohms.

Frequency (p. 120) $f = 1/T$ where $f$ represents the frequency (the number of cycles completed in one second) and $T$ is the time taken to complete one cycle (the period).